D0145635

Problems in Group Theory

John D. Dixon
Carleton University, Ottawa

Dover Publications, Inc.
Mineola, New York

Bibliographical Note

This Dover edition, first published in 1973 and reprinted in 2007, is an unabridged, corrected republication of the work originally published in 1967 by Blaisdell Publishing Company, Waltham, Massachusetts.

International Standard Book Number: 0-486-45916-0

Manufactured in the United States of America
Dover Publications, Inc., 31 East 2nd Street, Mineola, N.Y. 11501

Preface

Group theory is a basic branch of algebra. The simple definition of a group, as expressed through the group axioms, together with the way that group theory interpenetrates many other fields in mathematics, make it a good introduction to an advanced study of many areas of mathematics. In the century and a half since the foundations were laid, many great mathematicians have become interested in the study of group theory because its elementary nature scarcely conceals its many deep and tantalizing problems. Like number theory, group theory has many problems which may be stated in simple terms but whose solutions lie very deep.

Recent activity in group theory has resulted in the solution of several long-standing problems. Even a complete classification of all finite simple groups (the basic units from which all finite groups are built) now appears possible, in a large part due to the work of the mathematician John G. Thompson. On one hand, there has been the discovery and development of new tools applicable to old problems, and on the other, we have discovered how the older methods can be applied to new and interesting problems. The survey articles of Brauer [27] and Wielandt and Huppert [66] illustrate these developments.

To study group theory seriously, you have to read and understand the work of other mathematicians and then develop this understanding by solving suitable problems. Often it is difficult to find good problems which are challenging and yet accessible. This book is intended to provide such problems, with their solutions, from various branches of group theory. The problems are chosen and arranged so that they will be both challenging and accessible, and working through them should develop skill and understanding of the material. If you are unable to solve a given problem, then a study of its solution will prepare you for the later problems. Most of the problems are from research papers published since 1950, and there are many references to the original sources to help you find your way to the present frontiers of the subject. The references are meant to be suggestive rather than complete, but there are good bibliographies in Kurosh [2] (up to 1956), Curtis and Reiner [16] (on group representations), and the forthcoming book by Huppert [15] (on finite groups). The *Mathematical Reviews* and *Zentralblatt für Mathematik* give a continuing survey of current literature.

By the nature of the book, the focus of interest is as much on techniques used as on the results themselves, and this has partly dictated the arrangement of the material. This is especially true of the few standard theorems which appear as problems. The choice of problems has been influenced by my own preferences, and I have not attempted to touch all aspects of the subject. Some branches, such as the theory of infinite abelian groups and the study of free groups, are not represented because they require somewhat more specialized techniques than other parts of the subject. At the same time, I believe that the selection is sufficiently wide so that anyone studying group theory will find much that is of interest and of importance to him in these problems.

John D. Dixon

Acknowledgments

Much of what I know about group theory I learned from the following mathematicians at the California Institute of Technology: Everett Dade, Marshall Hall, Jr., and Olga Taussky-Todd. Perhaps they may recognize their respective influences in this book. I am also indebted to the California Institute of Technology for the use of its excellent library, as well as the University of New South Wales where part of this book was written.

This book owes much to the constructive criticisms of Garrett Birkhoff and Daniel Gorenstein who read an earlier version, and to Vlastimil Dlab whose careful checking and frank comments have eliminated a number of errors and inelegancies. I am grateful to these mathematicians and to Blaisdell Publishing Company for agreeing to publish this book.

This book is dedicated with affection to Hans and Hanna Schwerdtfeger for their early and persistent encouragement in all things mathematical.

J. D. D.

Contents

List of Symbols

Introduction

The following are prerequisites in group theory: a knowledge of the elementary properties of subgroups, the basic homomorphism laws, and the properties of finite direct products (such as might be covered in any introductory course in group theory). In algebra and elementary number theory, the prerequisite is a first course at the level of Birkhoff and MacLane's *Survey of Modern Algebra* or Herstein's *Topics in Algebra*. A good knowledge of linear algebra and matrix theory is necessary in Chapters 7, 10, and 11. A "naive" approach is taken towards set theory, and the axiom of choice and Zorn's lemma are used freely.

It is assumed that the student is presently studying or taking a course in advanced group theory at the level of the books by Hall [1], Kurosh [2], Rotman [3], Schenkman [4], and Scott [5]. Each chapter of problems begins with pertinent definitions and a list of propositions; for the proofs of the propositions there is reference to the books mentioned above. The chapters should be tackled successively, with the exception of Chapters 10 and 11, which are independent of the results in Chapters 7, 8, and 9. At the end of some chapters there are problems which are prefaced by an asterisk. These are sometimes (but not always) more difficult than the others and could be omitted at a first reading. Complete solutions of these problems are not included in the text, but references show where such solutions may be found. A few problems which require the verification of some simple property ("Show . . .") also have their solutions omitted. The following outline of basic definitions and notations is intended to refresh the memory of the reader and to clarify those notations which will be used without comment throughout the book.

NOTATION

We use the standard symbols for set operations and, in particular, denote the empty set by $\varnothing$. If f is a function of a set A into a set B, then we write $f(a)$ or a^f to denote the image of the element $a \in A$, and correspondingly use $f(S)$ or S^f to denote the image of a subset S of A. In particular, when we are dealing with groups of mappings we shall always use the right-hand notation a^f, and so the composite mapping fg will mean the mapping f followed by the mapping g. The *restriction* $f|S$ of f to a subset S of A is the function from S to B defined by $(f|S)(s) = f(s)$ for all $s \in S$.

Let a, b, and m be integers with $m \geqslant 1$. We write $a \mid b$ (respectively, $a \nmid b$) to mean that a does (respectively, does not) divide b, and write $a \equiv b \pmod{m}$ to denote that a is congruent to b modulo m. The set of all integers is partitioned into m disjoint residue classes (mod m), and these classes form a ring called "the ring of integers (mod m)" with a naturally defined addition and multiplication. If m is a prime, then this ring is a field. The *Euler ϕ-function* $\phi(m)$ denotes the number of positive integers $\leqslant m$ which are relatively prime to m, and it is well known that

$$\phi(m) = m \prod_{p|m} \left(1 - \frac{1}{p}\right).$$

The symbol $\binom{n}{k}$ denotes the binomial coefficient $n!/k!(n-k)!$.

The symbol $[\alpha_{ij}]$ will denote the matrix (of given dimensions) whose (i,j)th entry is α_{ij}. If $\mathbf{u}$ is a square matrix with complex entries, then $\det \mathbf{u}$ and $\operatorname{tr} \mathbf{u}$ denote the determinant and trace of $\mathbf{u}$, respectively. We use $\mathbf{1}$ (or $\mathbf{1}_r$) for the unit matrix of degree r.

In the following let G be a group, H a subgroup of G, and S a subset of G. (Except when otherwise specified, all groups will be written using the multiplicative notation.) Then $|S|$ will denote the order of S, and $|G:H|$ will denote the index of H in G. If G is a finite group, then H is a *Hall* subgroup of G if $|G:H|$ is relatively prime to $|H|$. By $\langle S \rangle$ we denote the subgroup of G generated by S; in particular, $\langle x \rangle$ is the cyclic subgroup generated by an element $x \in G$. In this context, S may be defined by certain conditions; for example, $\langle x \in G \mid x^2 = 1 \rangle$ is the subgroup generated by all elements of order 2. If $G = \langle S \rangle$, then S is a *set of generators* for G, and G is *finitely generated* if it has a finite set of generators. A group may sometimes be defined in terms of a set of generators satisfying a set of relations; we refer to the texts mentioned earlier for the details. (For example, see Kurosh [2], Section 18.) We use 1 to denote both the identity element of G and the subgroup $\langle 1 \rangle$; 1 is the *trivial* element (or subgroup). We say that H is a *proper* subgroup of G if $H \neq G$. An element $x \in G$ is a *p-element* if its order is a power of p; G is a *p-group* if all of its elements are p-elements, and it is an *elementary abelian p-group* if all its nontrivial elements have order p and the group is abelian.

For any $x \in G$, $x^{-1}Sx$ is called a *conjugate* of S in G; and the *normal closure* S^G of S in G is the smallest normal subgroup of G containing S; that is, $S^G = \langle x^{-1}Sx \mid x \in G \rangle$. The *centralizer* $\{x \in H \mid xy = yx$ for all $y \in S\}$ of S in H is denoted by $C(S;H)$; and the *normalizer* of S in H is $N(S;H) = \{x \in H \mid x^{-1}Sx = S\}$. The *center* of H is $Z(H) = C(H;H)$; and H is *abelian* if $Z(H) = H$. The set S is called a *normal* subset of G if G is the normalizer of S. A nontrivial group G is called *simple* if it has no normal subgroups except 1 and G. If N is a normal subgroup of G, then G/N denotes the *factor* (or *quotient*) *group*.

If F is a group, then a mapping θ of F into G is a *homomorphism* if $(xy)^\theta = x^\theta y^\theta$ for all $x,y \in F$. The *kernel* of θ is the normal subgroup $\ker \theta = \{x \in F \mid x^\theta = 1\}$. A homomorphism θ is an *isomorphism* if $\ker \theta = 1$ and θ maps F *onto* G; in this case we write $F \simeq G$. An isomorphism of the group G onto itself is called an *automorphism;* and if $S^\alpha = S$ for each automorphism α of G, then S is called a *characteristic* subset of G.

If A and B are groups, then the set of all ordered pairs $(a,b) (a \in A, b \in B)$ is a group under the multiplication

$$(a,b)(a',b') = (aa',bb') \quad (a,a' \in A;\, b,b' \in B).$$

This group is the (external) *direct product* of A by B, and we shall denote it by $A \times B$. If A and B are normal subgroups of a group G with $G = AB$ and $A \cap B = 1$, then G is the (internal) *direct product* of A by B. In this case G is isomorphic to the external direct product, and we shall use the same notation: $G = A \times B$. When dealing with additive groups we speak of *direct sums*, and we write $\dot{+}$ in place of $\times$. The concept of direct product generalizes easily to any finite set of groups.

When we apply the terms *maximal* and *minimal* to various sets of subgroups of G, we always consider the sets as ordered by set inclusion. For example, a maximal normal subgroup is a (proper) normal subgroup of G not contained in any other proper normal subgroup of G. In particular, the terms maximal subgroup and minimal subgroup mean maximal (proper) subgroup and minimal (nontrivial) subgroup, respectively.

Problems

1 Subgroups

The following results are assumed. For their proofs see Hall [1], Sections 1.5 to 1.6 and 2.1 to 2.3; Kurosh [2], Chapter 3; Rotman [3], Chapter 2; Schenkman [4], Sections 1.1 to 1.3; or Scott [5], Sections 1.6 to 1.7, 2.2 to 2.3, and 3.1 to 3.3.

1.T.1. If H is a subgroup of finite index in a group G, and K is a subgroup of G containing H, then K is of finite index in G, and

$$|G:H| = |G:K|\,|K:H|.$$

1.T.2. Let A and B be subgroups of a group G. If B is of finite index in G, then $A \cap B$ is a subgroup of finite index in A, and $|A:A \cap B| \leqslant |G:B|$. Equality holds if and only if $G = AB$.

In particular, if $|G:A|$ is also finite, then $|G:A \cap B| \leqslant |G:A|\,|G:B|$ with equality if and only if $G = AB$.

1.T.3. If S is a subset of a group G, then S has h conjugates $x^{-1}Sx$ ($x \in G$) in G if and only if $|G:N(S;G)| = h$. In the case that S consists of a single element s, then $N(S;G) = C(s;G)$, and s has $| G:C(s;G) |$ conjugates in G.

1.T.4. If N is a normal subgroup of a group G, then in the quotient group G/N the subgroup H/N is normal if and only if the subgroup H is normal in G. In the latter case, $G/H \simeq G/N \big/ H/N$.

1.T.5. If N and K are subgroups of a group G, and N is normal in G, then NK is a subgroup of G, and $NK/N \simeq K/(N \cap K)$.

1.T.6. If S is a subset of a group G, then S is a subgroup of G if and only if, for all $x,y \in S$, we have $x^{-1} \in S$ and $xy \in S$.

Problems

1.1. What is the order of a group G generated by two elements x and y subject only to the relations $x^3 = y^2 = (xy)^2 = 1$? List the subgroups of G.

1.2. What is the order of a group G generated by elements x and y subject only to the relations $x^3 = y^2 = (xy)^3 = 1$?

1.3. What is the order of a group G generated by elements x and y subject only to the relations $xy^2 = y^3x$ and $yx^3 = x^2y$?

1.4. Let Q be the group (under ordinary matrix multiplication) generated by the complex matrices

$$\mathbf{a} = \begin{bmatrix} 0 & 1 \\ -1 & 0 \end{bmatrix} \quad \text{and} \quad \mathbf{b} = \begin{bmatrix} 0 & i \\ i & 0 \end{bmatrix}$$

where $i^2 = -1$. Show that (a) Q is a nonabelian group of order 8; (b) each subgroup of Q is normal in Q; (c) Q is isomorphic to a group generated by elements x,y,z, and u subject only to the relations

$$x^2 = y^2 = z^2 = u, \qquad u^2 = 1, \qquad xy = z, \qquad yz = x, \qquad \text{and} \qquad zx = y.$$

(Q is called the *quaternion group.*)

1.5. Let T be the multiplicative group generated by the real matrices

$$\mathbf{u} = \begin{bmatrix} 0 & 1 \\ -1 & 0 \end{bmatrix} \quad \text{and} \quad \mathbf{v} = \begin{bmatrix} 0 & 1 \\ 1 & 0 \end{bmatrix}.$$

Then T is a nonabelian group of order 8, but T is not isomorphic to the group Q defined in Problem 1.4.

1.6. Let D_k be the group generated by elements a and b subject only to the relations

$$a^2 = b^k = (ab)^2 = 1 \qquad (k \geqslant 3).$$

(a) Show that D_k is isomorphic to the multiplicative group generated by the complex matrices

$$\mathbf{x} = \begin{bmatrix} 0 & 1 \\ 1 & 0 \end{bmatrix} \quad \text{and} \quad \mathbf{y} = \begin{bmatrix} \xi & 0 \\ 0 & \xi^{-1} \end{bmatrix}$$

where $\xi = \exp(2\pi i/k)$. (b) Show that D_k has order $2k$, and that D_4 is isomorphic to the group T defined in Problem 1.5. (c) List the conjugacy classes of D_k. (The groups D_k are called the *dihedral groups*.)

1.7. If H is a subgroup of finite index in a group G, then H contains a subgroup N which is of finite index and normal in G.

1.8. If A and B are subgroups of finite index in a group G, and $| G:A |$ and $| G:B |$ are relatively prime, then $G = AB$.

1.9. If H is a proper subgroup of a finite group G, then the set union $\bigcup_{x \in G} x^{-1}Hx$ is not the whole of G.

1.10. If H is a proper subgroup of finite index in a group G (possibly infinite), then there is some element $x \in G$ which is not in any subgroup conjugate to H in G. (Compare this with the result quoted in Problem *1.50.)

1.11. Let G be a group in which each proper subgroup is contained in a maximal subgroup of finite index in G. If every two maximal subgroups of G are conjugate in G, then G is a cyclic group.

1.12. Let S be a finite normal subset of a group G such that for some integer $n > 0$, $s^n = 1$ for all $s \in S$. Then every element of the group $H = \langle S \rangle$ may be written as a product of not more than $(n - 1) | S |$ elements of S. In particular, H is a finite group.

1.13. If H is a Hall subgroup of a finite group G, and K is another subgroup of the same order, then $N(H;K) = H \cap K$.

1.14. If N is a normal subset of a group G, then its centralizer $C(N;G)$ is a normal subgroup of G. If N is a characteristic subset of G (that is, each automorphism of G maps N into itself), then $C(N;G)$ is a characteristic subgroup of G.

1.15. Let N be a subgroup of the center $Z(G)$ of a group G. Then N is a normal subgroup of G. Moreover, if G/N is a cyclic group, then G is abelian.

1.16. A group G in which each element $x \neq 1$ has order 2 is abelian.

1.17. If N is a normal subgroup of finite index in a group G, and H is a subgroup of finite order in G, and if $|G:N|$ is relatively prime to $|H|$, then $H \subseteq N$.

1.18. If N is a normal subgroup of finite order in a group G, and H is a subgroup of finite index in G, and if $|G:H|$ is relatively prime to $|N|$, then $N \subseteq H$.

1.19. If M and N are normal subgroups of a group G, then $G/(M \cap N)$ is isomorphic to a subgroup of the (external) direct product $G/M \times G/N$.

1.20. Let x be an element of finite order n in a group G, and let $n = p_1^{k_1}p_2^{k_2} \cdots p_s^{k_s}$ where $p_1, p_2, \ldots, p_s$ are distinct primes. Then $x = x_1x_2 \ldots x_s$, where x_i is a p_i-element of order $p_i^{k_i}$ and $x_i = x^{m_i}$ for some integer m_i $(i = 1, 2, \ldots, s)$. Furthermore, if $x = y_1y_2 \ldots y_s$, where for $i,j = 1, 2, \ldots, s$ we have that y_i is a p_i-element and $y_iy_j = y_jy_i$, then $y_i = x_i$.

1.21. Let G be a group in which each subgroup has a finite number of conjugates in G. Then, for each subgroup H of G, $H \cap x^{-1}Hx$ is a subgroup of finite index in H for each $x \in G$.

1.22. Let G be a group in which each subgroup has a finite number of conjugates in G. Then every subgroup H of G contains a normal subgroup N of G such that N has finite index in H.

1.23. In any group G, the subset F of all elements which have only a finite number of conjugates in G is a characteristic subgroup.

A subgroup H of a group G will be called *special* if, for each pair of elements $x,y \in G$ with $x \notin H$, there is a unique $u \in H$ such that $y^{-1}xy = u^{-1}xu$.

1.24. If H is a special subgroup of a group G, and x is an element of G not in H, then $G = C(x;G)H$ and $C(x;G) \cap H = 1$.

1.25. A special subgroup H of a group G is normal in G.

1.26. If H is a special subgroup of a group G, and H is finite, then each coset $Hx \neq H$ $(x \in G)$ is a single class of elements conjugate in G.

1.27. If H is a special subgroup of a group G, then H contains all *commutators* $x^{-1}y^{-1}xy$ $(x,y \in G)$. In particular, if H is finite, then H is precisely the set of all such commutators, and so no group has more than one finite special subgroup.

1.28. If H is a special subgroup of a group G, then for each element x of G not in H the subgroup $U = C(x;G)$ is abelian.

1.29. With the notation of Problem 1.28, we have

$$U \cap y^{-1}Uy = 1 \qquad \text{for each} \qquad y \in G,\ y \notin U.$$

(A group which contains a proper subgroup U with this property is called a *Frobenius group*.)

A (finite or countably infinite) sequence of groups $G_0, G_1, \ldots,$ is called a *chain* if either of the following conditions hold: (a) for each $n \geqslant 1$, G_{n-1} is a subgroup of G_n; (b) for each $n \geqslant 1$, G_n is a subgroup of G_{n-1}. In case (a) the chain is *ascending*, and in case (b) the chain is *descending*. A chain is called *proper* if its terms are all different.

1.30. If $G_0, G_1, \ldots,$ is an ascending chain of groups, then an operation may be defined in a unique way in the set union

$$G = \bigcup_{n=0}^{\infty} G_n,$$

so that G is a group with each G_n as a subgroup. Moreover, if each G_n is abelian, then G is an abelian group.

1.31. Using the notation of Problem 1.30, let $H_0, H_1, \ldots,$ be a second ascending chain such that, for each $n \geqslant 0$, H_n is a normal subgroup of G_n. Then

$$H = \bigcup_{n=0}^{\infty} H_n$$

is a normal subgroup of G.

1.32. Let G be a finitely generated group. Then every proper ascending chain of subgroups of G whose set union equals G is finite.

A group G satisfies the *maximal* (respectively, *minimal*) *condition on subgroups* if each nonempty set of subgroups of G contains a subgroup M not contained in (respectively, not containing) any other subgroup of the set.

1.33. A group G satisfies the maximal condition on subgroups if each proper ascending chain of subgroups in G is finite.

1.34. A group G which satisfies the maximal condition on subgroups has the property that each proper ascending chain of subgroups of G is finite. [Converse of Problem 1.33.]

1.35. A group G satisfies the minimal condition on subgroups if and only if every proper descending chain of subgroups of G is finite.

1.36. A group G satisfies the maximal condition on subgroups if and only if each subgroup of G is finitely generated.

1.37. If a group G satisfies the maximal (respectively, minimal) condition on subgroups, then each factor group and each subgroup of G satisfies the same condition.

1.38. If A, B, and C are subgroups of a group G, and $A \subseteq C$, then $AB \cap C = A(B \cap C)$. [*Note:* AB is not necessarily a subgroup of G.]

1.39. Let A, B, and K be subgroups of a group G. If $A \subseteq B$, and we have $A \cap K = B \cap K$, and $AK = BK$, then $A = B$.

1.40. If N is a normal subgroup of a group G, and both N and G/N satisfy the maximal (respectively, minimal) condition on subgroups, then G satisfies the same condition. [Converse of Problem 1.37.]

A group G satisfies the *maximal condition on normal subgroups* if every nonempty set of normal subgroups of G contains a subgroup M which is not contained in any other subgroup in the set.

1.41. Let G be a group which satisfies the maximal condition on normal subgroups. If N is a normal subgroup of G, and G/N is isomorphic to G, then $N = 1$. (A group with this latter property is called a *Hopf* group.)

1.42. Let G be a finite group of order n with a subgroup H of order m. If $H \cap x^{-1}Hx = 1$ for all x in G not in H, then there are exactly $(n/m) - 1$ elements in G which do not lie in any conjugate of H.

1.43. If A and B are abelian subgroups of a group G, then $A \cap B$ is a normal subgroup of $\langle A,B \rangle$.

1.44. A finite noncyclic group G in which every proper subgroup is abelian possesses a normal subgroup N distinct from G and 1.

1.45. If G is a finite group such that, for each abelian subgroup A, $N(A;G) = C(A;G)$, then G is abelian.

1.46. Let G be a noncyclic group which is generated by two elements x and y each of order 2. Then G has a normal cyclic subgroup C of index 2 in G. If G is finite and of order $2k$, then G is isomorphic to the dihedral group D_k defined in Problem 1.6.

1.47. Let G be the multiplicative group generated by the real matrices

$$\mathbf{x} = \begin{bmatrix} 2 & 0 \\ 0 & 1 \end{bmatrix} \quad \text{and} \quad \mathbf{y} = \begin{bmatrix} 1 & 1 \\ 0 & 1 \end{bmatrix}.$$

Then G is a finitely generated group which does not satisfy either the minimal or the maximal condition on subgroups.

1.48. For each prime p, we define $Z(p^\infty)$ to be the set of all complex numbers z such that $z^{p^k} = 1$ for some integer $k \geqslant 0$. Show that (a) $Z(p^\infty)$ is an abelian group under multiplication; (b) the only proper subgroups of $Z(p^\infty)$ are the cyclic groups

$$T_n = \{z \in Z(p^\infty) \mid z^{p^n} = 1\} \qquad (n = 0, 1, \ldots);$$

(c) $Z(p^\infty)$ has no maximal subgroups and is not finitely generated; (d) $Z(p^\infty)/T_n$ is isomorphic to $Z(p^\infty)$ for each integer $n \geqslant 0$. (The group $Z(p^\infty)$ is called the *quasicyclic* or *Prüfer* p-group.)

1.49. Let G be the set of all matrices of the form

$$\begin{bmatrix} 2^k & p(x) \\ 0 & 1 \end{bmatrix}$$

where $k = 0, \pm 1, \pm 2, \ldots$, and $p(x)$ is any polynomial in x with rational coefficients. Show that (a) G is a group under matrix multiplication; (b) there exists a subgroup H of G and an element $\mathbf{u} \in G$ such that the left coset $\mathbf{u}H$ contains an infinite number of right cosets of H; (c) the chain $H \subset \mathbf{u}H\mathbf{u}^{-1} \subset \mathbf{u}^2H\mathbf{u}^{-2} \subset \cdots$ is an infinite proper ascending chain.

***1.50.** For each integer $r \geqslant 0$, there exists a number $\beta(r)$ such that any finite group with exactly r conjugacy classes has order at most $\beta(r)$. In particular, we may take $\beta(1) = 1$, $\beta(2) = 2$, $\beta(3) = 6$, $\beta(4) = 12$, and $\beta(5) = 60$. (The case for infinite groups is entirely different. Any infinite group in which each nontrivial element has infinite order may be embedded in a group with just two conjugacy classes: $\{1\}$ and the set of all nontrivial elements. See [6] or [2], Section 53.)

2 Permutation Groups

A *permutation* x of a set Ω is a one-to-one mapping of Ω onto itself. For each $\alpha \in \Omega$, we write α^x to denote the image of α under x. (The elements of Ω will be called *letters*.) The set S_Ω of all permutations of Ω is a group (called the *symmetric group* on Ω) under the product defined by the ordinary composition of mappings. If Ω has order n, and $\bar{\Omega} = \{1, 2, \ldots, n\}$, then S_Ω is isomorphic to $S_{\bar{\Omega}}$ under an obvious mapping. We denote the latter group by S_n and note that it has order $n!$. A subgroup G of S_n is called a permutation group of *degree* n. The set of all permutations in S_n which leave the letter n fixed is a subgroup which has a natural isomorphism with S_{n-1}, and we shall consider S_{n-1} as a subgroup of S_n under this identification. In particular, S_0 will denote the unit group.

The following results will be assumed. For their proofs see Hall [1], Sections 5.1 to 5.4; Kurosh [2], Sections 4 and 9; Rotman [3], Chapter 3; Schenkman [4], Section 3.7; or Scott [5], Section 1.3.

2.T.1. Each permutation of a finite set Ω may be written as a product of disjoint *cycles:* $x = (\alpha_1\alpha_2 \cdots \alpha_k)(\beta_1\beta_2 \cdots \beta_l) \cdots$, where, for example, the permutation $(\alpha_1\alpha_2 \cdots \alpha_k)$ maps $\alpha_1 \to \alpha_2$, $\alpha_2 \to \alpha_3$, $\cdots$, $\alpha_k \to \alpha_1$, and leaves all other letters fixed. These cycles commute, so the order in which they are written does not matter. Otherwise the representation for x in this form is unique except for cyclic permutations of the letters within the cycles. The *length* of a cycle is the number of letters appearing in it. The order of x, as an element of the group S_Ω, is the least common multiple of the lengths of its cycles.

2.T.2. A cycle of length 2 is a *transposition*. Each permutation of a finite set may be written as a product of transpositions (which, in general, are not disjoint). This may be done in many ways for a given permutation x, but the number of transpositions used is either always even or always odd. In the respective cases, x is called an *even* or an *odd* permutation. A cycle of length k is even or odd depending on whether $k + 1$ is even or odd.

2.T.3. The set of all even permutations in S_n is a characteristic subgroup A_n of index 2 in S_n, and A_n is called the *alternating group* of degree n.

2.T.4. If x and y are two permutations of a set Ω, and $x = (\xi_1\xi_2 \cdots \xi_k)$, then $y^{-1}xy = (\xi_1^y\xi_2^y \cdots \xi_k^y)$. [*Note:* The product xy denotes the result of first applying the mapping x and then the mapping y.] Thus, two permutations are conjugate in S_Ω if and only if they have the same cycle structure (in the sense of 2.T.1).

2.T.5. For $n \geqslant 5$, the only normal subgroup of S_n apart from S_n and 1 is the alternating group A_n. Furthermore, A_n is a simple group.

PROBLEMS

2.1. The alternating group A_4 has no subgroup of order 6.

2.2. Find the conjugacy classes of the symmetric group S_5 and the alternating group A_5 (using 2.T.4). Hence, show that A_5 is the only normal subgroup of S_5 (apart from 1 and S_5), and that A_5 is simple. (This gives a proof of a special case of 2.T.5.)

2.3. Find the normalizer N of the cyclic group $\langle(1\ 2 \ldots n)\rangle$ in the symmetric group S_n $(n \geqslant 1)$.

2.4. (a) Let $x = (1\ 2\ 3\ 4\ 5\ 6\ 7\ 8\ 9\ 10\ 11)$ and $y = (5\ 6\ 4\ 10)(11\ 8\ 3\ 7)$ be elements of the symmetric group S_{11}. Show that $M_{11} = \langle x, y\rangle$ has order $8 \cdot 9 \cdot 10 \cdot 11$. (b) Let $z = (1\ 12)(2\ 11)(3\ 6)(4\ 8)(5\ 9)(7\ 10)$. Show that $M_{12} =$

$\langle x, y, z\rangle$ is a subgroup of order 8·9·10·11·12 in S_{12}. (Both these groups are known to be simple groups. They were first discovered by Mathieu in 1860.)

2.5. For $n \geqslant 2$, the $n - 1$ transpositions (12), (13), . . . ,(1n) generate S_n.

2.6. For $n \geqslant 3$, the $n - 2$ 3-cycles (123), (124), . . . ,(12n) generate A_n.

2.7. A permutation group G of degree n which contains an odd permutation has a normal subgroup of index 2.

2.8. The group

$$S = \bigcup_{n=1}^{\infty} S_n$$

(see Problem 1.30) contains exactly one normal subgroup A distinct from S and 1. Furthermore, A is of index 2 in S.

Let G be an arbitrary group, and let Ω be a set for which, for each $\alpha \in \Omega$ and each $x \in G$, we have defined an element $\alpha^x \in \Omega$ with the properties: (a) the mapping $\bar{x}:\alpha \rightarrow \alpha^x$ is a permutation of the set Ω for each $x \in G$; and (b) $\bar{x}\bar{y} = \overline{xy}$ for all $x,y \in G$. Then, for each $\alpha \in \Omega$, the set $\alpha^G = \{\alpha^x \mid x \in G\} \subseteq \Omega$ is called the *orbit* (or transitivity set) of α, and the number of letters which α^G contains is the *length* of the orbit. The set $G_\alpha = \{x \in G \mid \alpha^x = \alpha\} \subseteq G$ is called the *stabilizer* (or stability subgroup) of α (see Wielandt [7]).

2.9. Two orbits, α^G and β^G $(\alpha,\beta \in \Omega)$, are identical whenever they have a letter in common. Thus, Ω is a set union of disjoint orbits.

2.10. The stabilizer G_α is a subgroup of G for each $\alpha \in \Omega$.

2.11. If $\beta = \alpha^x$ for some $x \in G$ and $\alpha,\beta \in \Omega$, then $x^{-1}G_\alpha x = G_\beta$. Thus, letters in Ω which belong to the same orbit have conjugate stabilizers in G.

2.12. If G is a finite group, then $|G| = |G_\alpha|\,|\alpha^G|$ for each $\alpha \in \Omega$.

2.13. If A and B are subgroups of a group G, then G is a union of disjoint *double cosets* of the form AxB $(x \in G)$. If A and B are finite subgroups, then the order of a double coset AxB is $|A|\,|B:x^{-1}Ax \cap B|$. [*Hint:* Consider the group B acting on the set of right cosets Ax $(x \in G)$.]

2.14. The mapping $x \rightarrow \bar{x}$ $(x \in G)$, where $\bar{x}$ is defined in (a) above defines a homomorphism of G into S_Ω. The kernel of the homomorphism is $\bigcap_{\alpha \in \Omega} G_\alpha$.

2.15. Let Ω be the set of all elements of a group G. We define $a^x = ax$ $(a, ax \in \Omega;\ x \in G)$. Then the homomorphism defined in Problem 2.14 is an isomorphism (which is called the *regular representation* of G).

2.16. If G is a group of order $2^k m$ (where m is odd), and G has a cyclic subgroup of order 2^k, then G has a normal subgroup of index 2^k. [*Hint:* Use Problems 2.7 and 2.15.]

2.17. Let H be a subgroup of a group G. We define Ω to be the set of all right cosets Ha $(a \in G)$, and define $(Ha)^x = Hax$ $(Ha, Hax \in \Omega;\ x \in G)$. Then the stabilizer of Ha is $a^{-1}Ha$, and the kernel K of the homomorphism defined in Problem 2.14 is $\bigcap_{a \in G} a^{-1}Ha$, which is the largest subgroup of H normal in G.

2.18. If a group G contains a subgroup H of index n, then it contains a normal subgroup K lying in H such that $|G:K|$ is finite and divides $n!$ (compare with Problem 1.7).

2.19. If G is a finite group of order m, and p is the smallest prime which divides m, then any subgroup H of index p in G is normal in G.

2.20. Let $n \geqslant 5$. Then the only proper subgroup of index less than n in the symmetric group S_n is the alternating group A_n (of index 2).

2.21. Let S be a finite set of order $p^r m$, where p is a prime (which may divide m). Let Ω be the set of all subsets consisting of p^r elements of S. Then $|\Omega| \equiv m \pmod{pm}$.

2.22. Let G be a finite group of order $p^r m$, where p is a prime (which may divide m), and let n_r denote the number of subgroups of order p^r in G. Then $n_r \equiv 1 \pmod{p}$, and hence $n_r \geqslant 1$. In particular, we have:

Sylow's theorem. If G is a finite group and p^s is the highest power of p which divides $|G|$, then G has at least one subgroup of order p^s, and the number of such subgroups is congruent to 1 (mod p). (These subgroups are called *Sylow p-groups* of G.)

2.23. If G is a finite group and P is a Sylow p-group of G, then any other p-subgroup Q of G is conjugate to some subgroup of P. In particular, all

Sylow p-groups of G are conjugate to one another in G, and every p-subgroup of G is contained in some Sylow p-group. Thus, G has $|G:N(P;G)|$ Sylow p-groups, and this latter number is congruent to 1 (mod p) by Problem 2.22. [*Hint:* Use Problem 2.13.]

2.24. How many Sylow subgroups (corresponding to the different primes) do nonabelian groups of orders 21 and 39, respectively, have?

2.25. There is no simple group of order 56.

2.26. There is no simple noncyclic group of order $2^m p^n$, when $m = 1, 2$, or 3 and p is an odd prime. (This is a particular case of a deep theorem of Burnside which states: a noncyclic group of order $p^m q^n$ (where p and q are primes) is not simple. The easiest proof of the latter theorem requires character theory.)

2.27. If G is a group of order 6, then either G is cyclic or G is isomorphic to the symmetric group S_3. [*Hint:* Use Problem 2.17.]

2.28. If a group G of order 12 has no element of order 2 in its center, then G is isomorphic to the alternating group A_4.

A permutation group G on a set Ω is called *transitive* if it has only one orbit, namely, Ω. Otherwise, G is called *intransitive*. A nonempty subset Γ of Ω is called a *block* (or set of imprimitivity) for G if, for each $x \in G$, either $\Gamma^x \cap \Gamma = \varnothing$ (the empty set) or $\Gamma^x = \Gamma$. All subsets of Ω which consist of a single letter and the subset Ω itself are blocks, and these blocks are called *trivial* blocks. We say that G is *primitive* if it has no nontrivial blocks; otherwise, it is *imprimitive*. We note that, for any nonempty set Ω, the group S_Ω is both transitive and primitive. A transitive group is called *regular* on Ω if no letter $\alpha \in \Omega$ is left fixed by a nontrivial element of G, that is, if $G_\alpha = 1$ for each $\alpha \in \Omega$.

2.29. For $n \geqslant 2$, $n - 2$ transpositions cannot generate a transitive group of degree n (compare with Problem 2.5).

2.30. If G is a transitive permutation group on a set Ω, then the stabilizers G_α $(\alpha \in \Omega)$ are all conjugate to one another in G. Thus, no stabilizer contains a nontrivial normal subgroup of G. Moreover, if $|\Omega| = n$, then $|G:G_\alpha| = n$, and so n divides $|G|$.

2.31. If G is a transitive group on a set Ω, then G is primitive if and only if the stabilizers G_α $(\alpha \in \Omega)$, which are conjugate by Problem 2.30, are maximal subgroups of G.

2.32. A primitive permutation group $G \neq 1$ is transitive.

2.33. A nontrivial normal subgroup of a primitive permutation group G is transitive.

2.34. A block Γ of a transitive group G of degree n has order dividing n. Thus, a transitive permutation group of prime degree p is primitive.

If G is a permutation group on a set Ω with orbits Ω_λ, then we denote by $G|\Omega_\lambda$ the group of permutations of Ω_λ given by the mappings in G restricted to the set Ω_λ.

2.35. $G|\Omega_\lambda \simeq G/N$, where

$$N = \bigcap_{\alpha \in \Omega_\lambda} G_\alpha.$$

2.36. If G is a permutation group on a set Ω with a finite number of orbits $\Omega_1, \Omega_2, \ldots, \Omega_s$, then G is isomorphic to a subgroup of the (external) direct product $G|\Omega_1 \times G|\Omega_2 \times \cdots \times G|\Omega_s$.

2.37. Let G be a permutation group of order p^k and degree n, where $n < p^2$. Then G is an elementary abelian p-group.

2.38. A regular group G of degree n has order n.

2.39. The image of the regular representation of a group G (see Problem 2.15) is a regular permutation group.

2.40. A finite transitive abelian group is regular, and so its order is equal to its degree.

2.41. The order of an abelian permutation group of degree n is not greater than $3^{n/3}$.

2.42. If G is a transitive group of degree n, then each element $x \neq 1$ in the center $Z(G)$ moves every letter. In particular, $|Z(G)| \leqslant n$.

2.43. Let G be a transitive permutation group on a set Ω, and let N be a normal subgroup of G. If Ω is of order n, and N has exactly k orbits $\Omega_1, \Omega_2, \ldots, \Omega_k$ in Ω, then (a) there exists an element $x_i \in G$ such that $\Omega_1^{x_i} = \Omega_i$ for $i = 1, 2, \ldots, k$; (b) the order of each orbit Ω_i is n/k; and (c) $N|\Omega_1 \simeq N|\Omega_i$ for $i = 2, 3, \ldots, k$.

2.44. A regular permutation group G of finite degree is primitive if and only if it has prime order.

2.45. If G is a primitive permutation group on a set Ω, then either G has prime order or, for each pair of distinct elements α and β in Ω, $G = \langle G_\alpha, G_\beta \rangle$.

2.46. Let G be a primitive permutation group on a set Ω. If G_γ is an abelian group for some $\gamma \in \Omega$, then $G_\alpha \cap G_\beta = 1$ for all $\beta \neq \alpha$ in Ω.

2.47. If G is a primitive permutation group of prime degree p, and for two stabilizers G_α and G_β we have $G_\alpha \cap G_\beta = 1$, then the Sylow p-group of G is normal in G.

2.48. If G is a primitive permutation group of prime degree p, and for two stabilizers G_α and G_β we have $G_\alpha \cap G_\beta = 1$, then each stabilizer G_γ is abelian.

2.49. If G is a transitive permutation group of degree n, and G has a normal p-subgroup $P \neq 1$, then $p \mid n$.

2.50. If G is a primitive permutation group of degree n and its center $Z(G)$ is nontrivial, then G is a cyclic group of prime order p, and $p = n$.

2.51. If a permutation group G contains a minimal normal subgroup N which is both transitive and abelian, then G is primitive.

2.52. Let G be a permutation group of degree n. If, for some integer k with $4 \leqslant 2k < n$, we have $|G| \geqslant (n-k)!k$, then G contains a permutation $x \neq 1$ which leaves fixed at least $n - 2k$ letters.

2.53. Every finite p-group P has a center $Z(P) \neq 1$. [*Hint:* Consider the set Ω of all elements of P, and the mapping $a^x = x^{-1}ax$ $(a, x^{-1}ax \in \Omega; x \in G)$.]

2.54. Let G be a group of order pq, where p and q are primes. If $p < q$ and $p \nmid q - 1$, then G is abelian.

2.55. Let G be a group of order n, and n be relatively prime to $\phi(n)$, where ϕ is the Euler ϕ-function. Then G is abelian. [*Hint:* Use Problem 1.44.]

2.56. Let p be a prime. Then every group G of order p^2 is abelian.

2.57. Find the order of the subgroup V of S_7 generated by the permutations (1 2 3 4 5 6 7) and (2 6)(3 4). Show that V has a subgroup P of order 7 whose normal closure P^V equals V. Hence prove that V is simple.

2.58. Any noncyclic simple group G whose order is at most 100 is of order 60.

***2.59.** Any simple group of order 60 is isomorphic to the alternating group A_5.

***2.60.** Describe all transitive permutation groups which are isomorphic to the symmetric group S_4. [*Hint:* Use Problem 2.17.]

***2.61.** List all the primitive subgroups of the symmetric groups S_5 and S_6.

***2.62.** Let G be a primitive subgroup of the symmetric group S_n, and let p be a prime $\leqslant n - 3$. If G contains an element x which is a p-cycle, then G contains the alternating group A_n. [*Hint:* Use Problem 2.33. First solve the problem when $p = 2$ or 3.]

***2.63.** Let x be any nontrivial element of the symmetric group S_n. If $n \neq 4$, then there exists an element $y \in S_n$ such that $S_n = \langle x,y \rangle$.

***2.64.** A Sylow p-group of an infinite group G is defined to be a subgroup of G which is maximal in the set of all p-subgroups of G. This agrees with the definition for finite groups because of the result of Problem 2.23. Unfortunately, the analogues of the results of Problems 2.22 and 2.23 are not necessarily true in infinite groups. For example, let G be the group generated by elements a, b, and c subject only to the relations

$$a^4 = 1, \qquad a^{-1}ba = ab^{-1}a^{-1} = c, \qquad \text{and} \qquad bc = cb.$$

Show that G has two classes of conjugate Sylow 2-groups of order 4, and a third class of conjugate Sylow 2-groups of order 2.

3 Automorphisms and Finitely Generated Abelian Groups

An *automorphism* of a group G is an isomorphism of G onto itself. The set of all automorphisms of G forms a group Aut G under the usual composition of mappings. The special automorphisms $\bar{u}$ ($u \in G$) defined by $\bar{u}: x \to u^{-1}xu$ ($x \in G$) are called *inner automorphisms*. We write automorphisms as mappings on the right; a product $\alpha\beta$ in Aut G means first α, then β.

The following results are assumed. For their proofs see Hall [1], Sections 6.1 and 3.1 to 3.2; Kurosh [2], Sections 12, 6, and 20; Rotman [3], Chapters 4 and 7; Schenkman [4], Sections 1.4, 2.3, and 3.2; or Scott [5], Sections 2.4 and 5.1.

3.T.1. The set of all inner automorphisms of a group G forms a normal subgroup of Aut G, and this subgroup is isomorphic to $G/Z(G)$.

3.T.2. For each integer $n \geqslant 1$, there is a cyclic group of order n (for example, the multiplicative group of all nth roots of 1 in the field of complex numbers). All cyclic groups of order n are isomorphic. If x is a generator of a cyclic group of order n, then x^k is also a generator if and only if k is relatively prime to n.

3.T.3. Every infinite cyclic group is isomorphic to the additive group of integers. If x is a generator of an infinite cyclic group, then the only other generator is x^{-1}.

3.T.4. Every subgroup and every factor group of a cyclic group is also cyclic. If G is a finite cyclic group of order n, then it contains a unique subgroup of order d for each positive divisor d of n. If G is an infinite cyclic group, then, for each integer $d \geqslant 1$, G has a unique subgroup of index d. The only other subgroup of the infinite cyclic group is the identity group. In particular, each subgroup of a cyclic group is a characteristic subgroup because it is the only one of that index.

3.T.5. If G is an abelian group which can be generated by n elements, then G can be written as a direct product, $G = C_1 \times \cdots \times C_m$, where the C_i are cyclic subgroups of G and $m \leqslant n$.

3.T.6. A direct product of a finite number of finite cyclic groups whose orders are relatively prime in pairs is cyclic.

Problems

3.1. How many mutually nonisomorphic abelian groups have orders p^2, p^3, p^4, and p^2q^3, respectively, where p and q are distinct primes?

3.2. Describe the five nonisomorphic groups of order 12. [*Hint:* Use Problem 2.28.]

3.3. Describe the different nonabelian groups of order p^3, where p is a prime. [*Hint:* Use Problems 2.19 and 2.22 to show that each group of order p^3 has a normal subgroup of index p. Then consider the cases $p = 2$ and $p \neq 2$ separately.]

3.4. Let G be a finite abelian group. Then G may be written as a direct product of cyclic subgroups in the form

$$G = C_1 \times C_2 \times \cdots \times C_m,$$

where the order h_i of C_i divides the order h_{i+1} of C_{i+1} for $i = 1, 2, \ldots, m-1$, and m is the smallest possible number of generators for G.

3.5. Let G be an abelian group generated by n elements. Then each subgroup of G may be generated by $\leqslant n$ elements.

3.6. The group A of all rational numbers (under addition) cannot be written as a direct sum of two or more nontrivial subgroups.

3.7. If a group G is generated by a subset S, then S^α also generates G for any automorphism α of G.

3.8. The group of all automorphisms of a cyclic group C of order n is an abelian group of order $\phi(n)$ (the Euler ϕ-function). The group of all automorphisms of an infinite cyclic group has order 2.

3.9. Find the group of all automorphisms of the symmetric group S_3.

3.10. Each automorphism of the symmetric group S_4 is an inner automorphism, and hence Aut $S_4 \simeq S_4$.

3.11. Give an example of two nonisomorphic groups which have isomorphic automorphism groups.

3.12. If G is a finite noncyclic abelian group, then Aut G is not abelian.

3.13. Let G be a finite group whose automorphism group acts transitively on the set of nontrivial elements of G, that is, for any x and y different from 1 in G, there exists $\alpha \in$ Aut G such that $x^\alpha = y$. Then G is an elementary abelian p-group for some prime p.

3.14. If the mapping $x \to x^{-1}$ $(x \in G)$ is an automorphism of the group G, then G is abelian.

3.15. Let α be an automorphism of a finite group G which leaves only the identity of G fixed. Then $G = \{x^{-1}x^\alpha \mid x \in G\}$.

3.16. Let G be a finite group. If Aut G has an element α of order 2 which leaves only the identity of G fixed, then G is an abelian group of odd order.

3.17. If H is a subgroup of a group G, then the factor group

$$N(H;G)/C(H;G)$$

is isomorphic to a subgroup of Aut H.

3.18. If K is a normal cyclic subgroup of a group G, then $G/C(K;G)$ is an abelian group of finite order. In particular, if $|K| = n$, then $|G:C(K;G)|$ divides $\phi(n)$ (the Euler ϕ-function); and if K is infinite, then $|G:C(K;G)| =$ 1 or 2.

3.19. Let H be a normal subgroup of a finite group G. If P is a Sylow p-group of H, then $G = N(P;G)H$.

3.20. If H is a normal subgroup of a group G, and K is a characteristic subgroup of H, then K is a normal subgroup of G. If, moreover, H is a characteristic subgroup of G, then K is a characteristic subgroup of G.

3.21. If G is a finite group, and n is a positive integer relatively prime to the order of G, then, for each $x \in G$, there is a unique $y \in G$ such that $y^n = x$. In particular, if $y^n = z^n$ for two elements y and x in G, then $y = z$.

3.22. Let G be a finite group of order mn, where m is relatively prime to n. Let A be a normal abelian subgroup of order m, and let H and K be subgroups of order n in G. Then there is an isomorphism θ of H onto K such that $Ax = Ax^\theta$ $(x \in H;\ x^\theta \in K)$. Moreover, for some $c \in A$, $cxc^{-1} = x^\theta$ for all $x \in H$. Thus, H is conjugate to K in G.

3.23. If H is a Hall subgroup of a finite group G, then $N(H;G)$ is its own normalizer in G.

3.24. If P is a Sylow subgroup of a finite group G, and two normal subsets S and T of P are conjugate to one another in G, then S is conjugate to T in $N(P;G)$.

3.25. If P is a Sylow subgroup of a finite group G, and P is contained in the center of $N(P;G)$, then every element of P lies in a different conjugacy class of G.

3.26. If C is a class of conjugate elements of a group G, and α is an automorphism of G, then C^α is also a class of conjugate elements of G.

3.27. The set A of all automorphisms of a group G which map each class of conjugate elements of G into itself is a normal subgroup of Aut G.

3.28. If G is a finite group, then the order of the group A defined in Problem 3.27 is only divisible by those primes which also divide $|G|$.

3.29. Let G be a finite group with abelian Sylow p-groups. If P is a Sylow p-group, and G contains no normal p-subgroups except 1, then $P \cap x^{-1}Px = 1$ for some $x \in G$. [*Hint:* Consider the subgroup H generated by P and $x^{-1}Px$, when x is chosen so that $P \cap x^{-1}Px$ has smallest possible order.]

3.30. If G is a finite group with abelian Sylow p-groups of order p^n, and 1 is the only normal p-subgroup of G, then G has at least $p^n + 1$ Sylow p-groups, and $|G| \geqslant p^n(p^n + 1)$.

3.31. If G is a finite group with an abelian Sylow p-group P, and G has no normal p-subgroup except 1, then the permutation group of P operating on the right cosets of P in G (as defined in Problem 2.17) has at least one orbit on which the representation for P is faithful.

3.32. Let G be a group which can be generated by a set of k elements. Then, for any integer $n > 0$, the number of normal subgroups of G which have index $\leqslant n$ is at most $(n!)^k$. [*Hint:* Use Problem 2.17.]

3.33. Let G be a finitely generated group. If H is a subgroup of finite index in G, then there is a characteristic subgroup K of G which is a subgroup of finite index in H (compare with Problem 1.7).

3.34. Let G be an elementary abelian p-group of order p^n. Then Aut G is isomorphic to the (multiplicative) group of all nonsingular $n \times n$ matrices with entries in the field of integers (mod p).

3.35. If G is an elementary abelian p-group of order p^n, then Aut G has order $(p^n - 1)(p^n - p) \cdots (p^n - p^{n-1}) = p^{n(n-1)/2}d$, where p does not divide d.

3.36. If G is an elementary abelian p-group of order p^n, and $n \leqslant p$, then G has no automorphism of order p^k for $k > 1$. [*Hint:* In a field of characteristic p, $\xi^{p^k} - 1 = (\xi - 1)^{p^k}$.]

3.37. Let m be a positive integer, and define G as the set of all matrices of the form

$$\begin{bmatrix} 1 & b \\ 0 & a \end{bmatrix}$$

where a and b are elements of the ring of integers (mod m), and a is relatively prime to m. Show that (a) G is a group under matrix multiplication, and G has order $m\phi(m)$, where $\phi(m)$ is the Euler ϕ-function; (b) if m is an odd prime, then each automorphism of G is an inner automorphism; and (c) if $m = 8$, then the mapping of G into itself defined by

$$\begin{bmatrix} 1 & b \\ 0 & a \end{bmatrix}^\alpha = \begin{bmatrix} 1 & b + \frac{1}{2}(a^2 - 1) \\ 0 & a \end{bmatrix}$$

is an outer automorphism of G which maps each conjugacy class of G onto itself. (Example (c) is due to G. E. Wall.)

3.38. Give an example of a group G which can be generated by two elements such that G has a subgroup H which is not finitely generated (compare with Problem 3.5).

3.39. Find a finite group G with a normal subgroup H such that $|\operatorname{Aut} H| > |\operatorname{Aut} G|$.

***3.40.** The quaternion group Q defined in Problem 1.4 has its automorphism group isomorphic to the symmetric group S_4. [*Hint:* Show that Aut Q has a normal subgroup of order 4 which is its own centralizer.]

***3.41.** Let p be a prime, and let G be the set of all matrices

$$\begin{bmatrix} a & b \\ c & d \end{bmatrix}$$

with entries in the field of integers (mod p) such that $ad - bc \equiv 1 \pmod{p}$. Show that (a) G is a group of order $p(p^2 - 1)$; (b) if $p \geqslant 3$, then the center Z of G has order 2; (c) if $p \geqslant 5$, then any normal subgroup H of G which contains Z as well as an element of the above form with $a = d = 1$, $b = 0$, and $c \not\equiv 0 \pmod{p}$ is equal to G; (d) if $p \geqslant 5$, then G/Z is a simple group of order $\frac{1}{2}p(p^2 - 1)$; and (e) if $p = 7$, then G/Z is isomorphic to the group V of Problem 2.57.

4 Normal Series

Let G be a group. A finite sequence of subgroups

$$G = G_0 \supseteq G_1 \supseteq \cdots \supseteq G_k = A, \tag{4.1}$$

which begins with G and ends with a subgroup A, is called a *normal series from G to A* when, for each i with $1 \leqslant i \leqslant k$, G_i is a normal subgroup of G_{i-1}. If no two subgroups in the series are equal, then the normal series is *without repetitions*. The *length* of the series (4.1) is k, and the factor groups G_{i-1}/G_i are called its *normal factors*. If $A = 1$, then (4.1) is called a *normal series for G*.

In the case that each G_i is a maximal normal subgroup of G_{i-1}, the normal series (4.1) is called a *composition series*. In the case that each G_i is maximal in the set of normal subgroups of G properly contained in G_{i-1} (for each i), then (4.1) is called a *principal series* (or chief series). Finally, if k is the smallest integer with the property that there is a normal series of length k from G to A, then (4.1) is called a *minimal normal series*.

Let us suppose that a second normal series from G to A is given by

$$G = H_0 \supseteq H_1 \supseteq \cdots \supseteq H_l = A. \tag{4.2}$$

If each subgroup in (4.1) also occurs in (4.2), then (4.2) is called a *refinement* of (4.1). On the other hand, if $k = l$, then the set of normal factors of (4.2) is said to be isomorphic to the set of normal factors of (4.1) if, for some ordering $i_1, i_2, \ldots, i_k$ of $1, 2, \ldots, k$, we have

$$G_{s-1}/G_s \simeq H_{i_s-1}/H_{i_s} \qquad (s = 1, 2, \ldots, k).$$

We shall assume the following results. For their proofs see Hall [1], Section 8.4; Kurosh [2], Section 16; Rotman [3], Chapter 6; Schenkman [4], Section 3.6; or Scott [5], Sections 2.5 and 2.10.

4.T.1. Every finite group has at least one composition series and at least one principal series.

4.T.2. Given any two normal series (4.1) and (4.2) from G to A, these series have refinements of equal lengths with isomorphic sets of normal factors.

4.T.3. If there exists a composition (respectively, principal) series for G, then all other composition (respectively, principal) series for G have isomorphic sets of normal factors. (This is known as the *Jordan-Hölder theorem.*)

Problems

4.1. Let H and K be two groups, and let θ be a homomorphism of K into Aut H. Thus, if θ_x is the image of $x \in K$ under θ, then the mapping $\theta_x : u \to u^x$ $(u \in H)$ is an automorphism of H, and we have $(u^x)^y = u^{xy}$ for all $u \in H$ and all $x,y \in K$. Associated with θ, we define the *semidirect* (or normal) *product* S of H by K as the set of all pairs (x,u) $(x \in K, u \in H)$, with the multiplication rule

$$(x,u)(x',u') = (xx',u^{x'}u'),$$

for all $x,x' \in K$ and all $u,u' \in H$. Show that (a) S is a group; (b) the sets $K_0 = \{(x,1) \mid x \in K\}$ and $H_0 = \{(1,u) \mid u \in H\}$ are subgroups of S isomorphic to K and H, respectively; and (c) H_0 is a normal subgroup of S, and $S = H_0K_0$ with $H_0 \cap K_0 = 1$.

4.2. Let D_k $(k \geqslant 3)$ be the dihedral group defined in Problem 1.6. Then Aut D_k is isomorphic to a semidirect product of a group of order k by a group of order $\phi(k)$, the Euler ϕ-function.

4.3. Give an example of a group G with a composition series such that G has a subgroup H which has neither a composition series nor a principal series.

4.4. Find a group G with a composition series

$$G = G_0 \supset G_1 \supset \cdots \supset G_n = 1,$$

which has the property that we cannot select any subseries

$$G = G_0 \supset G_{i_1} \supset \cdots \supset G_{i_k} = 1$$

(with $0 < i_1 < \cdots < i_k = n$) which is a principal series.

4.5. Let S and A be the infinite permutation groups defined in Problem 2.8. Let T be the abelian subgroup of S generated by the transpositions $(2m - 1\ 2m)$ $(m = 1, 2, \ldots)$. For each $x \in A$ we define an automorphism $\theta_x : u \to u^x$ $(u \in T)$ of T by putting $(2m - 1\ 2m)^x = (2n - 1\ 2n)$ if the permutation x maps the letter m onto n. Show that (a) there is an associated semidirect product G of T by A, and (b) G has a principal series, but G has no composition series.

4.6. The Jordan-Hölder theorem no longer holds if we extend the definition of "principal series" to include all proper descending chains, $H_0, H_1, \ldots,$ of normal subgroups of G with the properties: (a) $H_0 = G$ and $\bigcap_{k=0}^{\infty} H_k = 1$, and (b) for each $n \geqslant 1$, the normal series

$$H_0 \supset H_1 \supset \cdots \supset H_n$$

has no proper refinement.

4.7. If G is a finite p-group, then all the normal factors of a principal series of G are cyclic groups of order p. [*Hint:* Use Problem 2.53.]

4.8. If H is a proper subgroup of a finite p-group G, then $N(H;G) \neq H$. In particular, each maximal subgroup of G is normal.

A group is said to satisfy the *normal chain condition* if the following two conditions hold. (a) Each descending proper chain of subgroups, $G = G_0 \supset G_1 \supset G_2 \supset \cdots$, in which G_i is a normal subgroup in G_{i-1} for $i = 1, 2, \ldots$, has finite length. (b) Each ascending proper chain of subgroups, $1 = H_0 \subset H_1 \subset H_2 \subset \cdots$, in which H_i is a normal subgroup of H_{i+1} for $i = 0, 1, \ldots$, has finite length.

4.9. If G is a group satisfying the normal chain condition, and H is a normal subgroup of G, then both H and G/H satisfy the normal chain condition.

4.10. If G is a group satisfying the normal chain condition, then every proper normal subgroup of G is contained in some maximal normal subgroup of G. Similarly, every nontrivial normal subgroup of G contains a minimal normal subgroup of G. In particular, G possesses both maximal and minimal normal subgroups.

4.11. A group G satisfies the normal chain condition if and only if it has a composition series.

4.12. If G is a group satisfying the normal chain condition, and K is a minimal normal subgroup of G, then either K is simple or K is a direct product of a finite number of simple subgroups. Furthermore, these simple subgroups are all conjugate in G.

4.13. If G is a group satisfying the normal chain condition, then G has a principal series. Each normal factor of a principal series of G is either a simple group or a direct product of a finite number of mutually isomorphic simple groups.

A subgroup H of a group G is called *subnormal* (or subinvariant, accessible, or finitely serial) in G if there exists a normal series from G to H. If H is subnormal in G, then we denote by $m(G,H)$ the length of a minimal normal series from G to H.

4.14. If A and B are subnormal subgroups of a group G, then $A \cap B$ is also subnormal.

4.15. If H is a subnormal subgroup of a group G, then there is a normal series from G to H of the form

$$G = H_0 \supset H_1 \supset \cdots \supset H_m = H,$$

where H_i is the normal closure $H^{H_{i-1}}$ of H in H_{i-1} for $i = 1, 2, \ldots, m$, and $m = m(G,H)$.

4.16. If H is a subnormal subgroup of a group G, and α is an automorphism of G, then H^α is also subnormal in G, and $m(G,H^\alpha) = m(G,H)$.

4.17. Let H be a subnormal subgroup of a group G, and let α be an automorphism of G which maps H onto itself. Then every subgroup in the normal series from G to H given in Problem 4.15 is mapped onto itself by α.

4.18. If A and B are subnormal subgroups of a group G, and $A \subseteq N(B;G)$, then $\langle A,B\rangle$, which equals AB by 1.T.5, is a subnormal subgroup of G.

4.19. Let H be a subgroup of a group G. If the set $\mathcal{S}$ of all subnormal subgroups of G contained in H has a maximal element M, then M is normal in H. [*Note:* $\mathcal{S}$ is not empty since it contains 1.]

4.20. If G is a group such that each nonempty set of subnormal subgroups of G contains a maximal element, then, for every pair of subnormal subgroups A and B of G, $\langle A,B \rangle$ is a subnormal subgroup of G. In particular, this hypothesis holds if G has a composition series.

4.21. If H is a Hall subgroup of a finite group G, and H is subnormal in G, then H is normal in G.

4.22. If A and B are finite subgroups of a group G, and A is subnormal in G, then $\langle A,B \rangle$ is a finite group.

4.23. If A and B are finite subnormal subgroups of a group G, and the order of A is relatively prime to the order of B, then $\langle A,B \rangle$ is the direct product of A and B.

4.24. If A and B are finite subnormal subgroups of a group G, then the subgroup $\langle A,B \rangle$ which they generate is also subnormal in G. [*Note:* In contrast to the results of Problems 4.18, 4.20, and 4.24 (also see [8]), Zassenhaus [9] gives an example (Example 23, page 235) of a group in which the union of two subnormal subgroups is not subnormal.]

5 Commutators and Derived Series

In a group G, we define the *commutator* of two elements x and y to be $[x,y] = x^{-1}y^{-1}xy = [y,x]^{-1}$. If A and B are subsets of G, then we define $[A,B] = \langle [a,b] \mid a \in A \text{ and } b \in B \rangle = [B,A]$. Note that $[A,B] = 1$ if and only if $A \subseteq C(B;G)$.

The sequence of subgroups of G defined by

$$G = G^{(0)} \qquad \text{and} \qquad G^{(i)} = [G^{(i-1)},G^{(i-1)}] \qquad \text{for } i = 1, 2, \ldots$$

is called the *derived series* of G. Clearly,

$$G = G^{(0)} \supseteq G^{(1)} \supseteq \cdots \supseteq G^{(k)} \tag{5.1}$$

is a normal series from G to $G^{(k)}$, and each subgroup in it is a characteristic subgroup of G. The subgroups $G^{(1)}$, $G^{(2)}$, . . . , also written G', G'', . . . , are called the first, second, . . . , *derived groups* (or commutator subgroups) of G. If $G^{(l)} = 1$, and $G^{(l-1)} \neq 1$, then G is said to be *solvable* of *length* l.

We shall assume the following results whose proofs may be found in Hall [1], Section 9.2; Kurosh [2], Section 57; Rotman [3], Chapter 6; Schenkman [4], Sections 3.1 and 7.1; or Scott [5], Sections 2.6 and 3.4.

5.T.1. In the derived series, the normal factors are all abelian groups. In fact, $G^{(i)}$ is equal to the intersection of all the normal subgroups N of $G^{(i-1)}$ for which $G^{(i-1)}/N$ is abelian. In particular, if N is a normal subgroup of G, then G/N is abelian if and only if $G' \subseteq N$.

5.T.2. If $G^{(k-1)} \neq 1$ in (5.1), then

$$G/G^{(k)} \supset G^{(1)}/G^{(k)} \supset \cdots \supset G^{(k)}/G^{(k)} = 1$$

is a normal series without repetitions. In fact, it is the derived series for $G/G^{(k)}$.

5.T.3. If N is a normal subgroup of the group G, then $(G/N)' = G'N/N$.

5.T.4. If G is solvable, then each subgroup and each factor group of G is also solvable of length not exceeding that of G. The direct product of a finite number of solvable groups is solvable.

PROBLEMS

5.1. Commutators in a group G satisfy the following identities:

(a) $[xy,z] = y^{-1}[x,z]y[y,z]$,

(b) $[x,yz] = [x,z]z^{-1}[x,y]z$,

(c) $y^{-1}[[x,y^{-1}],z]y \cdot z^{-1}[[y,z^{-1}],x]z \cdot x^{-1}[[z,x^{-1}],y]x = 1$,

for all x, y, and z in G.

5.2. If a group G is generated by a subset K, then the smallest normal subgroup of G which contains $[K,K]$ is G'.

5.3. If H, K, and L are normal subgroups of a group G, then

$$\text{(a)} \quad [[H,K],L] \subseteq [[K,L],H][[L,H],K]$$

and

$$\text{(b)} \quad [HK,L] = [H,L][K,L].$$

5.4. If H and K are subgroups of a group G, then the subgroup $[H,K]$ is normal in $\langle H,K \rangle$.

5.5. Let H be a subgroup of a group G. If $[H,G'] = 1$, then $[H',G] = 1$.

5.6. Let x and y be two elements of order m and n, respectively, in a group G. If x and y both commute with $[x,y]$, and d is the greatest common divisor of m and n, then $[x,y]^d = 1$.

5.7. If a group G is generated by two elements, and the identity $[[x,y],y] = 1$ is satisfied for all $x,y \in G$, then $G' \subseteq Z(G)$.

Let x and y be elements of a group G. We define inductively the following series of commutators in G:

$$_0[x,y] = y \qquad \text{and} \qquad _n[x,y] = [x, {}_{n-1}[x,y]] \qquad \text{for } n = 1, 2, \ldots$$

In particular, $_1[x,y] = [x,y]$.

5.8. Let u be an element of a group G. If, for some integer $n > 0$, $_n[x,u] = 1$ for all $x \in G$, then $_{n+1}[u,x] = 1$ for all $x \in G$.

5.9. Give an example of a (finite) group G with an element u with the property that $_2[u,x] = 1$ for all $x \in G$ but, for some $y \in G$, $_n[y,u] \neq 1$ for all integers $n > 0$.

5.10. There is no group G which has its derived group G' isomorphic to the symmetric group S_4. [*Hint:* Use Problem 3.10.]

5.11. Let G be a group generated by elements x and y subject only to the relations $x^2 = 1$ and $x^{-1}yx = y^{-1}$. Show that (a) G is isomorphic to a semidirect product of an infinite cyclic group by a group of order 2; (b) each subgroup $G_k = \langle x, y^{2^k} \rangle$ $(k = 0, 1, \ldots)$ is a subnormal subgroup of G, but

$$H = \bigcap_{k=1}^{\infty} G_k$$

is not subnormal in G (in contrast to Problem 4.14); and (c) the subgroups $A = \langle x \rangle$ and $B = \langle xy \rangle$ each have order 2, but the commutator group $[A,B]$ is infinite. (G is called the *infinite dihedral group.* Compare with Problem 1.6.)

5.12. If N is a normal subgroup of a group G, and $N \cap G' = 1$, then $N \subseteq Z(G)$.

5.13. If A is a subgroup of a group G, then the normal closure A^G of A in G equals $[G,A]A$.

5.14. If A is a subgroup of a group G, then we can define the following chain of subgroups in G:

$$A_0 = G \qquad \text{and} \qquad A_n = [A, A_{n-1}] \qquad \text{for } n = 1, 2, \ldots$$

Then A is subnormal in G with $m(G,A) = m$ if and only if $A_m \subseteq A$ and A_{m-1} is not contained in A.

5.15. If A and B are subgroups which are subnormal in a group G, then

$$m(G,A \cap B) \leqslant \max \{m(G,A),m(G,B)\}.$$

(Compare with Problem 4.14.)

5.16. Let A be a normal abelian subgroup of a group G. If $x \in G$ and $a \in A$ and, for some positive integers h and n, $x^h \in A$ and ${}_n[x,a] = 1$, then $[x,a]^{h^{n-1}} = 1$.

An element x of a group G is called a *nilelement* (or Engel element) of G if, for each $y \in G$, there is an integer n, depending on y, such that ${}_n[x,y] = 1$.

5.17. If x is a nilelement of a group G, then, for each normal subgroup H of G, $N(\langle x\rangle;H) \subseteq \langle x\rangle$ implies that $H \subseteq \langle x\rangle$.

5.18. If a finite group G contains more than one Sylow p-group, and a p-element x is a nilelement of G, then x lies in at least two Sylow p-groups of G.

5.19. Let G be a finite group with more than one Sylow p-group, and let D be a p-subgroup of greatest possible order such that D lies in two different Sylow p-groups of G. Then $N = N(D;G)$ has more than one Sylow p-group.

5.20. If G is a finite group in which all p-elements are nilelements, then G has a unique (normal) Sylow p-group.

5.21. Let G be a group whose center Z has index n in G. Then G has at most n^2 different commutators.

5.22. Let G be a group whose center Z has index n in G. Then, for all $x,y \in G$, $[x,y]^{n+1} = [x,y^2][y^{-1}xy,y]^{n-1}$.

5.23. If the center Z of a group G has index n in G, then each element of G' may be written as a product of $\leqslant n^3$ commutators.

5.24. If a group G has center Z of index n in G, then G' is a finite group; in fact, $|G'| \leqslant n^{2n^3}$.

5.25. If G is a group whose derived group G' is finite of order m, then each element $x \in G$ has at most m conjugates in G.

5.26. If G is a group which can be generated by k elements, and the derived group G' of G has order m, then the center Z of G has finite index in G. In fact, $|G:Z| \leqslant m^k$. [Partial converse to Problem 5.24.]

5.27. If x and y are elements of a group G, and x and y both commute with $z = [x,y^{-1}]$, then $(xy)^s = x^s y^s z^{s(s-1)/2}$ for $s = 1, 2, \ldots$

5.28. Let G be a finite p-group where p is an odd prime. If G has only one subgroup of order p, then G is cyclic.

5.29. Let G be the group, under matrix multiplication, generated by the complex matrices

$$u = \begin{bmatrix} \xi & 0 \\ 0 & \xi^{-1} \end{bmatrix} \quad \text{and} \quad v = \begin{bmatrix} 0 & -1 \\ 1 & 0 \end{bmatrix}$$

where $\xi^{2^n} = 1$ and $\xi^{2^{n-1}} \neq 1$ for fixed $n \geqslant 2$. Show that (a) G is a nonabelian group of order 2^{n+1}; (b) G has only one subgroup of order 2; and (c) if $n > 2$, then G has some subgroups which are not normal. (G is called the *generalized quaternion group* of order 2^{n+1}. For $n = 2$, G is the quaternion group defined in Problem 1.4.)

***5.30.** Let G be a nonabelian group of order p^n in which each subgroup is normal. Then $p = 2$, and G is isomorphic to the direct product of the quaternion group Q and some (possibly trivial) elementary abelian 2-group.

***5.31.** Describe all the nonsolvable groups of order at most 200.

***5.32.** Let G be a group, and define $G(k) = \langle x^k \mid x \in G \rangle$ for $k = 1, 2, \ldots$ Let m and n be positive integers whose greatest common divisor is d. If $G(m)$ and $G(n)$ are both abelian groups, then $G(d)$ is an abelian group. Moreover, if $G(m)$ and $G(n)$ are cyclic, then $G(d)$ is also cyclic.

6 Solvable and Nilpotent Groups

Let G be a group. The *lower central series* of G is defined to be the normal chain:

$$G = {}^0G \supseteq {}^1G \supseteq \cdots \supseteq {}^iG \supseteq \cdots, \tag{6.1}$$

where ${}^iG = [G, {}^{i-1}G]$ for $i = 1, 2, \ldots$† All the terms in the lower central series are characteristic subgroups of G. If, for some k, ${}^kG = 1$ and ${}^{k-1}G \neq 1$, then G is called *nilpotent* of *class* k.

The *upper central series* for G is defined by

$$1 = Z_0 \subseteq Z_1 \subseteq \cdots \subseteq Z_i \subseteq \cdots, \tag{6.2}$$

where Z_i is the characteristic subgroup of G defined by $Z_i/Z_{i-1} = Z(G/Z_{i-1})$ for $i = 1, 2, \ldots$ In particular, $Z_1 = Z(G)$.

---

† There are several different notations used for the lower central series. It should be noted that the enumeration is often different; thus our ${}^0G$, ${}^1G$, $\cdots$ are sometimes called the first, second, $\cdots$ terms of the lower central series.

The following results are assumed. For their proofs see Hall [1], Sections 9.2 and 10.2; Kurosh [2], Sections 57 and 62; Rotman [3], Chapter 6; Schenkman [4], Sections 6.1 and 7.1; or Scott [5], Section 6.4.

6.T.1. If G is a group with a normal subgroup K, and K is solvable of length m, and G/K is solvable of length n, then G is solvable of length $\leqslant m + n$.

6.T.2. The normal factors of the series (6.1) and (6.2) are all abelian. Thus, a nilpotent group of class k is solvable of length at most k.

6.T.3. If G is a nilpotent group, then each subgroup and each factor group of G is nilpotent of class not exceeding the class of G. A direct product of a finite number of nilpotent groups is nilpotent.

Problems

6.1. Give an example of a group G with a normal subgroup N such that G/N and N are both nilpotent but G is not nilpotent (compare with 6.T.1).

6.2. Let G be the generalized quaternion group of order 2^{n+1} defined in Problem 5.29. Show that $G'' = 1$, and that G is nilpotent of class n.

6.3. If, in the upper central series (6.2) of a group G, $Z_k = G$, then, in the lower central series (6.1), ${}^iG \subseteq Z_{k-i}$ for $i = 1, 2, \ldots, k$.

6.4. A group G is nilpotent of class k if and only if, in its upper central series (6.2), $Z_k = G$ and $Z_{k-1} \neq G$. Thus, for a nilpotent group, the upper and lower central series reach G and 1, respectively, after the same number of terms.

6.5. A finite p-group is nilpotent. [*Hint:* Use Problem 2.53.]

6.6. A finite group G with all its Sylow subgroups normal is nilpotent.

6.7. Let N be a minimal normal subgroup of a group G, and let K be a nilpotent normal subgroup of G. Then $N \subseteq C(K;G)$.

6.8. If G is a nilpotent group of class k, then each subgroup H is subnormal in G. In fact, we have $m(G,H) \leqslant k$.

6.9. In a nilpotent group, each maximal subgroup (if any exists) is a normal subgroup. Conversely, if each maximal subgroup of a finite group G is normal, then G is nilpotent.

6.10. In a finite nilpotent group, each Sylow subgroup is normal. [Converse of Problem 6.6.]

6.11. A minimal normal subgroup N of a finite solvable group G is an elementary abelian p-group for some prime p.

6.12. A maximal subgroup M of a finite solvable group G has its index equal to a power of some prime.

6.13. Each nontrivial normal subgroup N of a nilpotent group G has a nontrivial intersection with $Z(G)$. If G is finite, then each minimal normal subgroup of G is a subgroup of prime order in $Z(G)$.

6.14. Let G be a nilpotent group. Then the set of all p-elements of G forms a normal subgroup of G for each prime p. [*Hint:* Use Problem 4.22.]

A group G satisfies the *normalizer condition* if each proper subgroup H of G is a proper subgroup in its normalizer $N(H;G)$. It follows from Problem 6.8 that each nilpotent group satisfies the normalizer condition.

6.15. Let G be a group in which each subgroup is subnormal. If N is a normal subgroup of G, $Z(N) \neq 1$, and G/N is cyclic, then $Z(G) \neq 1$.

6.16. Let G be a group which satisfies the normalizer condition and the maximal condition on subgroups. Then each subgroup of G is subnormal.

6.17. A group G which satisfies the normalizer condition and the maximal condition on subgroups has a nontrivial center.

6.18. A group G which satisfies both the normalizer condition and the maximal condition on subgroups is nilpotent.

6.19. Let G be a group with lower and upper central series (6.1) and (6.2), respectively. Then, for $i = 1, 2, \ldots$, and $j \geqslant i$, we have $[Z_j, {}^{i-1}G] \subseteq Z_{j-i}$. In particular, $[Z_i, {}^{i-1}G] = 1$.

6.20. If G is a nilpotent group of class k, then $[{}^{i-1}G, {}^{k-i}G] = 1$ for $i = 1, 2, \ldots, k$.

6.21. If G is a nilpotent group of class k, and n is the greatest integer $\leqslant \frac{1}{2}k$, then ${}^nG$ is an abelian group.

6.22. If G is a nilpotent group of class k, then G is solvable of length l where $2^{l-1} \leqslant k$ (compare with Problem 6.2).

6.23. Let H be a normal subgroup of a group G. We define inductively the subgroups

$$H_0 = H \qquad \text{and} \qquad H_i = [G,H_{i-1}] \qquad \text{for } i = 1, 2, \ldots,$$

and the subgroups

$$U_n = \coprod_{i=0}^{n} [H_i, H_{n-i}] \qquad \text{for } n = 0, 1, 2, \ldots$$

Then $U_0, U_1, U_2, \ldots,$ is a sequence of normal subgroups of G such that $[U_n, G] \subseteq U_{n+1}$ for $n = 0, 1, 2, \ldots$

6.24. If H is a normal subgroup of a group G, and H and G/H' are nilpotent groups of classes k and m, respectively, then G is nilpotent of class $\leqslant (2^k - 1)m - 2^{k-1} + 1$.

6.25. If H and K are normal nilpotent subgroups of a group G, and $G = HK$, then

$${}^nG \subseteq {}^nH{}^nK({}^0H \cap {}^{n-1}K)({}^1H \cap {}^{n-2}K) \cdots ({}^{n-1}H \cap {}^0K) \qquad \text{for } n = 1, 2, \ldots$$

6.26. Let H and K be normal nilpotent subgroups of a group G, and the classes of H and K be h and k, respectively. If $G = HK$, then G is nilpotent, and the class of G is at most $h + k$.

6.27. If H is a maximal nilpotent subgroup of a group G, and $N = N(H;G)$, then N is its own normalizer in G.

Let G be a group. The *nilradical* (or Fitting subgroup) of G is the set $R(G)$ of elements x in G which lie in some normal nilpotent subgroup of G.

6.28. The nilradical of a group G is a characteristic subgroup of G. If G is finite, then $R(G)$ is a normal nilpotent subgroup of G. In any case, any subgroup H of G which can be generated by a finite number of elements from $R(G)$ is nilpotent.

6.29. All elements lying in the nilradical $R(G)$ of a group G are nilelements of G.

6.30. If G is a finite solvable group, then the nilradical $R(G)$ is just the set of all nilelements of G.

6.31. If, in the lower central series (6.1) for a group G, the factor group $G/{}^1G$ is a cyclic group, then ${}^iG = {}^1G$ for each integer $i \geqslant 1$. In particular, if G is a finite p-group with $|G| > p$, then $|G:G'| > p$.

6.32. Let H be a normal subgroup of a group G such that $H \subseteq G'$. If H' and H/H' are both cyclic groups, then $H' = 1$. In particular, in the derived series of a group G, two successive factor groups $G^{(k-1)}/G^{(k)}$ and $G^{(k)}/G^{(k+1)}$ cannot both be nontrivial cyclic groups when $k \geqslant 2$.

6.33. Let G be a nilpotent group with normal subgroups M and N. If N is a proper subgroup of M, then there is a normal subgroup K of G such that $N \subset K \subseteq M$ and the factor group K/N is cyclic.

6.34. Let G be a nilpotent group satisfying the maximal condition on normal subgroups. Then there exists a proper chain of normal subgroups of G of the form

$$G_0 = 1 \subset G_1 \subset \cdots \subset G_n = G,$$

where G_i/G_{i-1} is a cyclic group for $i = 1, \ldots, n$.

6.35. If G is a nilpotent group satisfying the maximal condition on normal subgroups, then each subgroup H of G is finitely generated (compare with Problem 1.36).

6.36. If A is a maximal normal abelian subgroup of a nilpotent group G, then $C(A;G) = A$. [*Note:* A direct application of Zorn's lemma shows that every group possesses a maximal normal abelian subgroup, possibly the group itself.]

6.37. If G is a nilpotent group, and N is a finite normal subgroup of order p^n in G, for some prime p, then $N \subseteq Z_n$, where Z_n is defined as in the upper central series (6.2).

6.38. Let A be a normal abelian p-subgroup of a group G, and let N be a normal cyclic subgroup of G contained in A. If $|A:N| = p$ and $[A,G] \subseteq N$ then, either A is a cyclic group, or $A = \langle N,a \rangle$ where a is an element of A with at most p conjugates in G.

6.39. Let G be a finite p-group. If G' has a cyclic center $Z(G')$, then G' is an abelian group.

6.40. If G is a finite p-group, and H is a nontrivial normal subgroup of G, then H contains a subgroup K which is normal in G with $| H:K | = p$.

6.41. If G is a finite p-group, and $| G':G'' | \leqslant p^2$, then G' is an abelian group.

6.42. Every finite group G contains a nilpotent subgroup K whose normal closure K^G equals G.

6.43. Let G be a transitive permutation group of degree n. If G is nilpotent, then each prime p which divides $| G |$ also divides n.

6.44. If G is a transitive permutation group of degree n, and G is a p-group, then the solvable length l of G satisfies $p^l \mid n$. [*Note:* It follows from Problem 2.30 that n is a power of p.]

6.45. If G is a nilpotent transitive permutation group of degree n, then the order of G divides $\prod_{p|n} p^{p^k-1/p-1}$, where k is the highest power to which p divides n, and the product is taken over all primes p which divide n.

6.46. If G is a nilpotent permutation group of degree n, then the order of G is at most 2^{n-1} (compare with Problem 2.41).

6.47. Let G be a group which has a normal subgroup A of index 2 such that A is isomorphic to the quasicyclic group $Z(2^\infty)$, defined in Problem 1.48, and $G = \langle b,A \mid b^{-1}ab = a^{-1}$ for all $a \in A$; $b^2 = 1\rangle$. (Thus, G is isomorphic to a semidirect product of $Z(2^\infty)$ by a group of order 2.) Show that (a) G is a 2-group which satisfies the normalizer condition; (b) ${}^iG = G'$ for all $i \geqslant 1$ where ${}^iG$ is defined in (6.1), hence G is not nilpotent; and (c) $\bigcup_{i=1}^{\infty} Z_i(G) = G'$ and $G'' = 1$. (This example is due to R. Baer.)

6.48. Let G be a nonabelian group of order p^m with an abelian subgroup of index p. Then G has either 1 or $p+1$ abelian subgroups of index p, and in the latter case $| G:Z(G) | = p^2$.

6.49. A group G which has order 16 and is nilpotent of class 3 has one, and only one, cyclic subgroup of index 2.

***6.50.** Describe all groups of order 2^n which have class $n-1$ ($n \geqslant 3$). [*Note:* It follows from Problem 6.31 that the index of the derived group of a finite 2-group is at least 4, and so $n - 1$ is the maximum possible class of a group of order 2^n.]

***6.51.** Up to isomorphism, there are exactly 14 different groups of order 16.

***6.52.** If G is a group of order $p^n q$, where p and q are primes, then G is solvable.

***6.53.** If G is a finite group in which every proper subgroup is nilpotent, then G is solvable.

***6.54.** Let G be a group in which each nontrivial element has order 3. Then G is nilpotent of class at most 3 (compare with Problem 1.16).

7 The Group Ring and Monomial Representations

Let $\mathcal{C}$ denote the field of complex numbers. Let G be a group, and consider the set R_G of all formal sums $\sum_{x \in G} \alpha_x x$ $(\alpha_x \in \mathcal{C})$ in which all but a finite number of coefficients α_x are zero. We define addition and multiplication in R_G by

$$(\sum_{x \in G} \alpha_x x) + (\sum_{x \in G} \beta_x x) = \sum_{x \in G} (\alpha_x + \beta_x)x$$

and

$$(\sum_{x \in G} \alpha_x x)(\sum_{x \in G} \beta_x x) = \sum_{x \in G} \gamma_x x,$$

where $\gamma_x = \sum_{z \in G} \alpha_{xz^{-1}} \beta_z$. (Note that γ_x is a finite sum of elements in $\mathcal{C}$ because β_z is zero for all but a finite number of $z \in G$.) An element $\sum_{x \in G} \alpha_x x$ in R_G, which, for some $u \in G$, has $\alpha_u = 1$ and $\alpha_x = 0$ for $x \neq u$, is written as u and is said to be an element of R_G lying in G. It is readily shown that R_G is an associative ring with unity element 1 (the identity of G), and that

R_G is commutative if and only if G is abelian. We call R_G the *group ring* of G (over $\mathbb{C}$).

We wish to consider matrices over R_G. Since R_G is generally noncommutative and possesses divisors of zero, the theory of such matrices is more difficult than the corresponding theory of matrices over a field. For this reason we consider only a very special class of matrices. The set of $n \times n$ *monomial matrices* over G is defined to consist of those $n \times n$ matrices which have precisely one nonzero entry in each row and each column with all nonzero entries lying in G. *Permutation matrices* are those monomial matrices for which each nonzero entry is 1; and a monomial matrix is called *diagonal* if its nonzero entries all lie on the main diagonal. A diagonal matrix is written $\operatorname{diag}(x_1, x_2, \ldots, x_n)$, where $x_1, x_2, \ldots, x_n$ are the successive diagonal entries.

The proof of the following properties is straightforward.

7.T.1. Every monomial matrix may be written as the (matrix) product $\mathbf{uv}$ of a permutation matrix $\mathbf{u}$ and a diagonal matrix $\mathbf{v}$.

7.T.2. The set $S(n)$ of all $n \times n$ permutation matrices forms a group, under matrix multiplication, which is isomorphic to the symmetric group S_n.

7.T.3. The diagonal matrices in the set of all $n \times n$ monomial matrices over G form a group $D(n,G)$. This group is isomorphic to the direct product $G \times G \times \cdots \times G$ (n times).

7.T.4. The set $M(n,G)$ of all $n \times n$ monomial matrices over G is a group in which $D(n,G)$ is a normal subgroup. Moreover,

$$M(n,G) = S(n)D(n,G) \qquad \text{and} \qquad S(n) \cap D(n,G) = 1.$$

7.T.5. If G is an abelian group, then we define the *determinant* ($\det \mathbf{a}$) of a monomial matrix $\mathbf{a}$ over G as the product of its nonzero entries. There is no ambiguity in the definition because G is abelian. Thus, $\det \mathbf{a}$ is an element of G, and for $\mathbf{a},\mathbf{b} \in M(n,G)$ we find $\det \mathbf{ab} = (\det \mathbf{a})(\det \mathbf{b})$.

Let G be a group with a subgroup H of finite index n in G. Let θ be a homomorphism of H into a group S. Then we define $\mathring{\theta}$ as a function of G into the group ring of S by

$$\mathring{\theta}(x) = \begin{cases} \theta(x) & \text{if } x \in H, \\ 0 & \text{otherwise.} \end{cases}$$

Let $r_1, r_2, \ldots, r_n$ be a set of left coset representatives for H in G. We shall now define the *monomial representation* θ^G of G induced from θ over the given set of coset representatives. For each $x \in G$, we define

$$\theta^G(x) = \begin{bmatrix} \dot{\theta}(r_1^{-1}xr_1) & \dot{\theta}(r_1^{-1}xr_2) & \cdots & \dot{\theta}(r_1^{-1}xr_n) \\ \dot{\theta}(r_2^{-1}xr_1) & \dot{\theta}(r_2^{-1}xr_2) & \cdots & \dot{\theta}(r_2^{-1}xr_n) \\ \cdot & & & \cdot \\ \cdot & & & \cdot \\ \cdot & & & \cdot \\ \dot{\theta}(r_n^{-1}xr_1) & \dot{\theta}(r_n^{-1}xr_2) & \cdots & \dot{\theta}(r_n^{-1}xr_n) \end{bmatrix} = [\dot{\theta}(r_i^{-1}xr_j)]. \quad (7.1)$$

The matrix $\theta^G(x)$ lies in $M(n,S)$. To prove this we have to show that for each r_i there is exactly one r_j and one r_k such that $r_i^{-1}xr_j \in H$ and $r_k^{-1}xr_i \in H$. In fact, r_j and r_k are the uniquely defined left coset representatives of the cosets $x^{-1}r_iH$ and xr_iH, respectively.

Problems

7.1. Let θ^G be the function of the group G into $M(n,S)$ as defined by (7.1). Then θ^G is a homomorphism of G into $M(n,S)$. If the kernel of θ is N, then the kernel of θ^G is $\bigcap_{x \in G} x^{-1}Nx$.

7.2. Using the same notation as above, let $s_1, s_2, \ldots, s_n$ be a second set of left coset representatives for H in G. Define θ_1^G to be the monomial representation of G induced from θ over this new set of coset representatives. Then there exists an $n \times n$ monomial matrix $\mathbf{u}$ over S such that $\mathbf{u}^{-1}\theta^G(x)\mathbf{u} = \theta_1^G(x)$ for all $x \in G$.

7.3. Let H be a subgroup of G, and put $\theta(x) = xH'$ for all $x \in H$. Then the mapping τ of G into H/H' defined by $\tau(x) = \det \theta^G(x)$ $(x \in G)$ is a homomorphism of G into H/H'. Moreover, τ is independent of the particular choice of left coset representatives for H in G. The mapping τ is called the *transfer* of G into H.

7.4. Let G be a group possessing a normal abelian Hall subgroup A of order m and index n in G. If H is a Hall subgroup of order n in G, and K is a subgroup of G such that n divides $|K|$, then, for some $x \in G$, $x^{-1}Hx \subseteq K$. [*Hint:* Use Problem 3.22.]

7.5. Let G be a group containing a normal abelian Hall subgroup A of order m and index n in G. Then there exists a subgroup U of order n in G. It follows that $G = UA$ and $U \cap A = 1$. [*Hint:* Use Problems 7.1 and 7.4.]

7.6. Let H be a normal Hall subgroup of a finite group G. Then H has a *complement* K in G, that is, a subgroup K of G such that $G = KH$ and $H \cap K = 1$. [*Hint:* Use Problem 3.19.]

7.7. Let H be a normal Hall subgroup of a finite group G. Then any two complements K and L of H in G are conjugate in G if either (a) H is solvable, or (b) G/H is solvable. [*Note:* Because H is a Hall subgroup, one of the groups H and G/H has odd order. Thus, from the theorem of Feit and Thompson [10], at least one of the conditions (a) and (b) is always true.]

7.8. If the center $Z(G)$ of a group G has index h in G, then the mapping $x \to x^h$ $(x \in G)$ defines a homomorphism of G into $Z(G)$. [*Hint:* Use Problem 7.3.]

7.9. If the center $Z(G)$ of a group G has index h in G, then each element of G' has an order dividing h (compare with Problem 5.24).

Let H be a subgroup of a finite group G. A set $R = \{r_1, r_2, \ldots, r_n\}$ of left coset representatives for H in G is called *exceptional* if, for each $u \in H$, $u^{-1}Ru = R$. It is clear that an exceptional set of left coset representatives is also a set of right coset representatives.

7.10. Let H be a subgroup of a finite group G. Then H has an exceptional set of coset representatives in G if and only if (*) for all $x \in G$ there exists $y \in HxH$ such that $y^{-1}Hy \cap H = C(y;H)$.

7.11. Let H be an abelian Hall subgroup of a finite group G. Then H has an exceptional set of coset representatives in G if and only if there is a normal subgroup K of G for which $G = HK$ and $H \cap K = 1$. (Such a subgroup K is called a *normal complement* of H in G.)

7.12. Let P be a Sylow p-group of a finite group G such that $C(P;G) = N(P;G)$. (In particular, P must be abelian.) Then P has a normal complement in G. [*Hint:* Use Problem 3.25.]

7.13. Let G be a finite group of order n, and let p be a prime dividing n such that n is relatively prime to $p - 1$. If the Sylow p-groups of G are cyclic, then any Sylow p-group of G has a normal complement in G. (This is a generalization of Problem 2.16.)

7.14. If G is a finite group which has all its Sylow subgroups cyclic, then $G'' = 1$. [*Hint:* Use Problem 6.32.]

7.15. If G is a noncyclic simple group of even order, then either 8 or 12 divides $|G|$. [*Hint:* Use Problems 7.12 and 3.17.]

7.16. There are no simple groups of order 300, 400, or 540.

7.17. If a noncyclic simple group G has odd order, then its order would be divisible by p^3, where p is the smallest prime dividing $|G|$.

7.18. There is no noncyclic simple group of odd order less than 1000. (It is proved in [10] that every group of odd order is solvable.)

7.19. Let G be a finite solvable group. If $|G| = mn$, where m and n are relatively prime, then G has a (Hall) subgroup of order m. Moreover, if K is a subgroup of G, and the order of K divides m, then $x^{-1}Kx \subseteq H$ for some $x \in G$.

7.20. Let G be a finite solvable group of order $n = p_1^{k_1} \cdots p_s^{k_s}$, where the p_i are distinct primes. Then G has a set of Sylow subgroups $P_1, \ldots, P_s$ such that P_i is a Sylow p_i-group and $P_iP_j = P_jP_i$ for all i and j.

7.21. Let G be a finite solvable group. Then G contains a nilpotent subgroup H such that $N(H;G) = H$, that is, H is *self-normalizing* in G.

7.22. Let G be a finite solvable group. Then any two self-normalizing nilpotent subgroups H and K of G are conjugate in G.

7.23. Let G be a finite group, and let H be a subgroup with the property that $x^{-1}Hx \cap H = 1$ for each $x \in G$ with $x \notin H$. Then H is a Hall subgroup of G.

7.24. Let G be a finite group possessing a maximal subgroup A which is abelian. Then the third derived group $G^{(3)}$ is 1.

7.25. Give an example of a group G of order mn with m relatively prime to n such that G has no Hall subgroup of order m (compare with Problem 7.19).

7.26. Let G be the group of all 5×5 monomial matrices over a cyclic group of order 5, and put $H = G'$. Show that (a) $|H| = 2^2 \cdot 3 \cdot 5^5$; (b) $H' = H$, although the center of H is nontrivial.

7.27. For any finite group B of order n, there exists a group C of order n^{n+1} such that C has a normal subgroup B_0 isomorphic to B and $B_0 \subseteq C'$. [*Hint:* Construct C as a monomial group over B.]

7.28. Construct an ascending chain of groups G_n ($n = 0, 1, \ldots$) with the following properties: (a) each group G_n is a finite p-group; (b) $G_n \subseteq G'_{n+1}$ for all $n \geqslant 0$; and (c) $G = \bigcup_{n=0}^{\infty} G_n$ is a p-group with $G' = G$. In particular, this shows that an infinite p-group need not be solvable. [*Hint:* Use Problem 7.27.]

8 Frattini Subgroup

The *Frattini subgroup* $\Phi(G)$ of a group G is defined to be the intersection of all maximal subgroups of G (if G has any maximal subgroups), and to be G itself (if G has no maximal subgroup). Clearly, $\Phi(G)$ is a characteristic subgroup of G.

The following results will be assumed. For their proofs see Hall [1], Section 10.4; Kurosh [2], Section 62; Schenkman [4], Section 1.7; or Scott [5], Section 7.3.

8.T.1. The Frattini subgroup of a nontrivial group G is the set of all elements x in G with the property that, whenever a set $K \cup \{x\}$ generates G, K generates G.

8.T.2. If G is a nilpotent group, then $G' \subseteq \Phi(G)$.

8.T.3. If G is a finite group, and $G' \subseteq \Phi(G)$, then G is nilpotent.

8.T.4. If G is a group with a finitely generated Frattini subgroup, then the only subgroup H of G such that $H\Phi(G) = G$ is $H = G$ (use 8.T.1).

PROBLEMS

8.1. Find the Frattini subgroups of the following groups: the cyclic groups; the symmetric groups S_n; the dihedral groups D_k defined in Problem 1.6; the generalized quaternion groups defined in Problem 5.29; and the quasicyclic groups $Z(p^\infty)$ defined in Problem 1.48.

8.2. Let H be a subgroup of a finitely generated group G. Then $H\Phi(G) = G$ implies that $H = G$.

8.3. Let G be a nilpotent group whose derived group G' is finitely generated. Then the only subgroup H of G such that $HG' = G$ is $H = G$.

8.4. If G is a finite group, and if the only subgroup H of G with the property $HG' = G$ is $H = G$, then G is nilpotent.

8.5. Let G be a group satisfying the maximal condition on subgroups. If H is a subgroup of G, and H is not normal in G, then, for some $x \in G$, the subgroup $x^{-1}Hx$ is not equal to any of the conjugates $u^{-1}Hu$ with $u \in \Phi(G)$.

8.6. If G is a finite group, then $\Phi(G)$ is nilpotent.

8.7. Let G be a group with a finitely generated Frattini subgroup Φ. If G/Φ can be generated by k elements and no fewer, then G can also be generated by k elements and no fewer.

8.8. If G is a finite group, and P is a nontrivial Sylow subgroup of G, then P is not contained in $\Phi(G)$.

8.9. Let G be a group satisfying the maximal condition on subgroups. Then, for each subnormal subgroup H of G, $\Phi(H) \subseteq \Phi(G)$.

8.10. If N is a normal subgroup of a group G, and $N \subseteq \Phi(G)$, then $\Phi(G/N) = \Phi(G)/N$.

8.11. Let $G = \langle x \rangle$ be an infinite cyclic group. Then $\Phi(G) = 1$, but there exist normal subgroups N of G such that $\Phi(G/N) \neq 1$.

8.12. Give an example of a group G such that $G/\Phi(G)$ is a finite non-trivial group but G is not finitely generated (compare with Problem 8.7).

8.13. Give an example of a finite group G which has a subgroup H such that $\Phi(H)$ is not contained in $\Phi(G)$ (compare with Problem 8.9).

8.14. It is shown in Problem 8.6 that every finite group has a nilpotent Frattini subgroup; however, there is no finite group G which has its Frattini subgroup F isomorphic to the quaternion group. [*Hint:* Putting $C = C(F;G)$, show that CF/C is contained in $\Phi(G/C)$ and then use Problem *3.40.]

A finite group G is called *p-closed* (for a prime p) if its Sylow p-group is a normal subgroup. In this case, the Sylow p-group is unique (by Problem 2.23). Thus, G is p-closed if and only if the product of any two p-elements in G is again a p-element.

8.15. If M is a subnormal subgroup of a finite group G, and M is maximal with respect to the properties that it is subnormal and p-closed, then M is normal in G.

8.16. Let M be a normal subgroup of a finite group G such that $\Phi(G) \subseteq M$. If $M/\Phi(G)$ is p-closed, then M is p-closed. [*Hint:* Use Problem 8.5.]

8.17. Let H be a subnormal subgroup of a finite group G. Then H is p-closed if $H/(H \cap \Phi(G))$ is p-closed.

8.18. If H is a subnormal subgroup of a finite group G, then H is nilpotent if and only if $H' \subseteq \Phi(G)$ (compare with 8.T.2 and 8.T.3).

8.19. If N is a normal subgroup of a group G, and G has a finitely generated subgroup U such that $N \subseteq \Phi(U)$, then $N \subseteq \Phi(G)$.

8.20. For any group G, $Z(G) \cap G' \subseteq \Phi(G)$.

8.21. If G is a direct product of two subgroups A and B, then

$$\Phi(G) \subseteq \Phi(A) \times \Phi(B).$$

8.22. If a finitely generated group G is the direct product of two subgroups A and B, then $\Phi(G) = \Phi(A) \times \Phi(B)$.

8.23. Let A and H be two subgroups of a group G such that $G = AH$. If A is a normal abelian subgroup of G, then $H \cap A$ is also normal in G.

8.24. Let A be a finite normal abelian subgroup of a group G such that $A \cap \Phi(G) = 1$. Then there is a subgroup K of G such that $G = KA$ and $K \cap A = 1$.

If N is a subgroup of a group G which is mapped into itself by an automorphism α of G, then we can define the *induced automorphism* $\bar{\alpha}$ of the factor group G/N by $\bar{\alpha}: Nx \to Nx^{\alpha}$ $(Nx \in G/N)$.

8.25. Let G be a finite group which can be generated by a set of k elements, and whose Frattini subgroup is Φ. Then $|\operatorname{Aut} G|$ divides $|\Phi|^k \, |\operatorname{Aut} G/\Phi|$. In particular, if $\alpha \in \operatorname{Aut} G$ leaves G/Φ elementwise fixed, then the order of α divides $|\Phi|^k$.

8.26. For any finite p-group G, $G/\Phi(G)$ is an elementary abelian p-group.

8.27. Let G be a finite group of order p^n with $|G:\Phi(G)| = p^k$. Then $|\operatorname{Aut} G|$ divides $p^{k(n-k)}(p^k - 1)(p^k - p) \cdots (p^k - p^{k-1}) = p^{k(2n-1-k)/2}d$, where $p \nmid d$. Hence, $|\operatorname{Aut} G|$ always divides $p^{n(n-1)/2}d$. [*Hint:* Use Problem 3.35.]

8.28. A finite p-group G of order p^n contains a normal abelian subgroup A whose order p^m satisfies $m(m+1) \geqslant 2n$. [*Hint:* Use Problem 6.36.]

8.29. If G is a finite solvable group, and $\Phi_1(G)$ denotes the intersection of all subgroups of prime index in G, then $G'' \subseteq \Phi_1(G)$. [*Hint:* Use Problem 3.18.]

8.30. Let N be a normal subgroup of a finite group G. Then there exists a subgroup H of G such that $G = HN$, and $H \cap N$ is a nilpotent group.

8.31. Let N be a normal subgroup of index n in a group G. Let A be the group of all automorphisms of G which leave each element of N fixed, and map each coset of N into itself. Then A is isomorphic to a subgroup of the direct product B of n copies of the center of N: $B = Z_1 \times Z_2 \times \cdots \times Z_n$ with each $Z_i \simeq Z(N)$. In particular, A is abelian.

8.32. Let H be a group isomorphic to the quasicyclic group $Z(5^{\infty})$ defined in Problem 1.48. Show that (a) there exists a sequence $k_0, k_1, k_2, \ldots,$ of integers such that $k_0 = 1$, $k_1 = 2$, $k_{n+1} \equiv k_n \pmod{5^n}$ and $k_{n+1}^2 \equiv -1$

(mod 5^{n+1}) for $n = 1, 2, \ldots$; and (b) the mapping α of H into itself defined by $x^\alpha = x^{k_n}$, when $x \in H$ has order 5^n, is an automorphism of order 4 for H.

Define G as a group which is isomorphic to a semidirect product of H by a cyclic group of order 4 such that

$$G = \langle u,H \mid u^4 = 1,\ u^{-1}xu = x^\alpha \quad \text{for all} \quad x \in H\rangle.$$

Show that (c) each maximal subgroup of G contains H; and (d) the Frattini subgroup $\Phi(G)$ of G equals $\langle u^2,H\rangle$ and, in particular, is not nilpotent (compare with Problem 8.6). (See [11].)

8.33. If A is a finite abelian group, then there exists a finite abelian group G such that $\Phi(G) \simeq A$ (compare with Problem 8.14).

9 Factorization

If A and B are proper subgroups of a group G, and $G = AB$, then G is called *factorizable*. If, in addition, $A \cap B = 1$, then B is called a *complement* of A in G.

If A and B are any two subgroups of a group G such that $AB = BA$, then these subgroups are said to *commute*. In particular, each normal subgroup N of G commutes with every subgroup of G.

The results of 1.T.2 and Problem 1.8 should be noted. In particular, 1.T.2 implies the following result.

9.T.1. If A and B are subgroups of a finite group G, then $|A||B| \leqslant |G||A \cap B|$. There is equality if and only if $G = AB = BA$.

PROBLEMS

9.1. Let G be the symmetric group S_5. Then G has a subgroup H which has two nonisomorphic complements in G.

9.2. For which integers $n \geqslant 3$ does the subgroup A_{n-1} of the alternating group A_n have a complement in A_n?

9.3. Construct a group G with two cyclic subgroups A and B such that $G = AB$ but neither A nor B is normal in G.

9.4. Give an example of a group G which has two subnormal subgroups which do not commute.

9.5. If A and B are subgroups of a group G, then A and B commute if and only if $AB = \langle A,B \rangle$.

9.6. If a group G is a product AB of two abelian subgroups A and B, then $Z(G) = (A \cap Z(G))(B \cap Z(G))$.

9.7. If G is a factorizable group, $G = AB$, then for any elements x and y in G we have $G = (x^{-1}Ax)(y^{-1}By)$.

9.8. If G is a factorizable group, $G = AB$, then A is not conjugate to B in G.

9.9. A finite group G is nilpotent if and only if each pair of maximal subgroups of G commute.

9.10. Let A be a subgroup of finite index in a group G. If A commutes with each of its conjugates in G, then A is subnormal in G.

9.11. If A and B are abelian subgroups of a group G, and $G = AB$, then every pair of commutators of the form $[a,b]$ $(a \in A; b \in B)$ commute. Hence $G'' = 1$.

9.12. If G is a factorizable group, $G = AB$, and $A \cap B$ contains a nontrivial normal subgroup of the group B, then the subgroup A contains a nontrivial normal subgroup of G.

9.13. If a group G is a product AB of an abelian subgroup B and any other subgroup A, then, either A contains a nontrivial normal subgroup of G, or $B \cap x^{-1}Ax = 1$ for all $x \in G$.

9.14. Let G be a finite group which is a product AB of abelian subgroups A and B with $A \neq B$. Then either A or B is contained in a proper normal subgroup N of G.

9.15. If a finite group G is a product AB of proper abelian subgroups A and B, then either A or B contains a nontrivial normal subgroup of G.

9.16. Let A and B be subgroups of relatively prime orders m and n, respectively, in a finite group G. If $G = AB$, and H is a normal subgroup of G, then $H = (A \cap H)(B \cap H)$.

9.17. Let A and B be subgroups of a group G such that $G = AB$ and the index of B in G is finite. If N is a finite normal subgroup of A, and the order of N is relatively prime to the index $|G:B|$, then the normal closure N^G of N in G lies in B. [*Hint:* Use Problem 1.18.]

9.18. Let G be a finite group, and let A, B, and C be solvable subgroups of G. If the indices $|G:A|$, $|G:B|$, and $|G:C|$ are relatively prime in pairs, then G is solvable.

The lattice of subgroups of a group G is called *modular* if the subgroups satisfy the following condition. For any subgroups A, B, and C of G with $B \subseteq A$:

$$\langle A \cap C, B\rangle = A \cap \langle B, C\rangle.$$

(See Hall [1], page 117, or Kurosh [2], Section 44.) A group whose lattice of subgroups is modular is called an *M-group.*

9.19. Let G be a group in which each pair of subgroups commute. Then G is an M-group satisfying the normalizer condition. [*Hint:* Use Problem 9.10.]

9.20. Every pair of subgroups of a group commute whenever it is true that every pair of cyclic subgroups of G commute.

9.21. If G is a group such that, for every pair of subgroups A and B,

$$(*) \qquad N(A;B) = A \cap B \qquad \text{implies that} \qquad B \subseteq A,$$

then each pair of subgroups of G commute.

9.22. If G is a group which satisfies the normalizer condition, then each subgroup H of G also satisfies the normalizer condition.

9.23. If G is an M-group which satisfies the normalizer condition, then every pair of subgroups of G commute. (This is the converse of Problem

9.19. A finite group satisfies the normalizer condition if and only if it is nilpotent—by Problem 6.9.)

9.24. Construct a group G with two subgroups A and B and a normal subgroup N such that $G = AB$ but $N \neq (N \cap A)(N \cap B)$. (Compare with Problems 9.6 and 9.16.)

***9.25.** Describe all finite groups with the property that *every* subgroup has a complement.

10 Linear Groups

Let $\mathcal{C}$ be the field of complex numbers, and let $\mathcal{V}$ be a vector space of dimension n over $\mathcal{C}$. Then the *general* (or full) *linear group* $GL(\mathcal{V})$ is the group of all invertible linear transformations of $\mathcal{V}$ into itself, under composition of mappings; and the *special linear group* $SL(\mathcal{V})$ is the subgroup of $GL(\mathcal{V})$ consisting of those linear transformations with determinant 1. (All linear transformations will be written as functions on the right.) A *scalar* is a linear transformation $\alpha 1$ ($\alpha \in \mathcal{C}$) where 1 is the identity on $\mathcal{V}$. A subgroup of $GL(\mathcal{V})$ is called a *linear group*.

If a basis is chosen for $\mathcal{V}$, then to each linear transformation of $\mathcal{V}$ into itself there corresponds a uniquely determined $n \times n$ matrix with entries in $\mathcal{C}$. This correspondence defines an isomorphism from $GL(\mathcal{V})$ onto the *general linear group* $GL(n)$ of all nonsingular $n \times n$ matrices over $\mathcal{C}$, under matrix multiplication; this isomorphism maps $SL(\mathcal{V})$ onto the *special linear group* $SL(n)$ consisting of the matrices in $GL(n)$ with determinant 1. Subgroups of $GL(n)$ will be called *matrix groups*, and the correspondence between linear groups and matrix groups will be used repeatedly in discussing their properties. We use boldface letters, $\mathbf{x}$, $\mathbf{y}$, . . . , to denote matrices, and write $[\xi_{ij}]$ to denote a matrix, of given dimensions, in which the (i,j)th entry is ξ_{ij}.

There are certain important subgroups of $GL(n)$. The (lower) *triangular group* $TL(n)$ consists of those matrices $[\xi_{ij}]$ in $GL(n)$ with $\xi_{ij} = 0$ for all $i < j$. The *special triangular group* $STL(n)$ consists of the matrices in $TL(n)$ with diagonal entries all 1. The *diagonal group* $D(n)$ consists of all diagonal matrices in $GL(n)$, and $S(n)$, which is isomorphic to S_n, consists of all permutation matrices in $GL(n)$. (A permutation matrix is one in which each row and each column has exactly one nonzero entry, and that entry is 1.) Finally, $M(n)$ is the *monomial group* consisting of all matrices in $GL(n)$ which have a single nonzero entry in each row and each column. We note that $D(n)$ is a normal subgroup of $M(n)$, and $M(n) = S(n)D(n)$ with $S(n) \cap D(n) = 1$ (compare with Chapter 7). If G is a subgroup of $GL(\mathcal{V})$ and, for a suitable basis of $\mathcal{V}$, G corresponds to a subgroup of $D(n)$ [$M(n)$ or $TL(n)$], then we say that G corresponds to a diagonal [monomial or triangular] group.

If $\mathcal{V}$ has dimension n, then elements and subgroups of $GL(\mathcal{V})$ are said to have *degree n*. If G is a subgroup of $GL(\mathcal{V})$, then a subspace $\mathcal{W}$ of $\mathcal{V}$ is an *invariant subspace* for G if $\mathcal{W}x = \mathcal{W}$ for all $x \in G$. In this case, the restrictions $x|\mathcal{W}$ are elements of $GL(\mathcal{W})$, and the group $\{x|\mathcal{W} \mid x \in G\}$ is denoted by $G|\mathcal{W}$ and is called the *component* of G on $\mathcal{W}$. We say that G is *irreducible* if it has no invariant subspace except $\mathcal{V}$ and 0 (the subspace consisting of the zero vector alone); otherwise, G is *reducible*. G is *completely reducible* if $\mathcal{V} = \mathcal{W}_1 \dotplus \cdots \dotplus \mathcal{W}_k$ (direct sum), where the $\mathcal{W}_i$ are minimal invariant subspaces for G; in particular, an irreducible group is completely reducible. The groups $G|\mathcal{W}_i$ $(i = 1, \ldots, k)$ are called the *irreducible components* of G.

The above terms may also be applied to subgroups of $GL(n)$ in the following way. We fix a basis for $\mathcal{V}$. Then to each subgroup G of $GL(\mathcal{V})$ there corresponds an isomorphic subgroup $\mathbf{G}$ of $GL(n)$. From elementary matrix theory, we know that for any other basis of $\mathcal{V}$ the corresponding subgroup will be some conjugate $\mathbf{c}^{-1}\mathbf{G}\mathbf{c}$ of $\mathbf{G}$ in $GL(n)$, where $\mathbf{c}$ in some sense describes the change of basis. If G is reducible, and $\mathcal{W}$ is an invariant subspace of dimension r, then we may choose a basis for $\mathcal{V}$ containing a basis for $\mathcal{W}$. Then the elements of the subgroup of $GL(n)$ corresponding to $\mathbf{G}$ will have the form

$$\begin{bmatrix} \mathbf{x}_1 & \mathbf{0} \\ \mathbf{x}_3 & \mathbf{x}_2 \end{bmatrix}$$

where $\mathbf{x}_1$, $\mathbf{x}_2$, and $\mathbf{x}_3$ are blocks of dimensions $r \times r$, $(n - r) \times (n - r)$, and $(n - r) \times r$, respectively. Thus, we say that a subgroup $\mathbf{G}$ of $GL(n)$ is *reducible* if it is similar, that is, conjugate in $GL(n)$, to a group of matrices of the above form for some r, $1 \leqslant r < n$. Similarly, **G** is *completely reducible* if for certain $r_1, \ldots, r_k$ it is similar to a group of matrices of the form

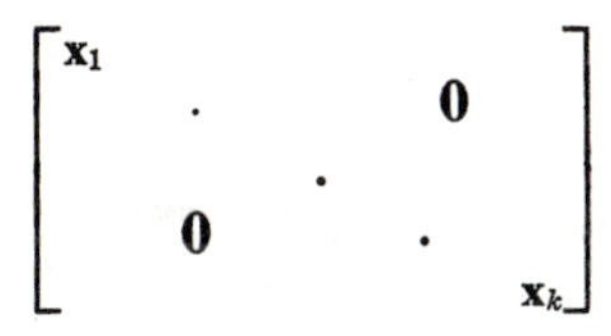

where the irreducible blocks $\mathbf{x}_i$ on the diagonal are $r_i \times r_i$ $(i = 1, \ldots, k)$.

For the proof of the following result see Hall [1], Section 16.3; Schenkman [4], Theorem IV.8.m; or Scott [5], Section 12.1.

10.T.1. If G is a finite linear group (over $\mathcal{C}$), then G is completely reducible. (This result is known as Maschke's Theorem.)

Note: In many cases, the results of this chapter are valid when the field involved is quite general. Sometimes, however, the properties "algebraically closed" and "of characteristic zero" which $\mathcal{C}$ possesses are crucial. Results like Problem 3.34 indicate why a generalization to fields other than the complex numbers is of interest in the study of abstract groups. For such generalizations the reader is referred to the literature quoted in the solutions to the problems of this chapter. Typical applications to abstract groups appear in [12]; [13]; [14]; and [15].

PROBLEMS

Note: Throughout this chapter, $\mathcal{V}$ will denote a finite dimensional vector space over the complex field $\mathcal{C}$.

10.1. Let G be an irreducible subgroup of $GL(\mathcal{V})$. If $c \in GL(\mathcal{V})$ centralizes G, then $c = \gamma 1$ for some $\gamma \in \mathcal{C}$. In particular, the irreducible components of a completely reducible abelian group $A \subseteq GL(\mathcal{V})$ are all of degree 1, and so A corresponds to a diagonal group for a suitable basis for $\mathcal{V}$. [*Hint:* Show that $\mathcal{V}(c - \alpha 1)$ is an invariant subspace for G for each $\alpha \in \mathcal{C}$.]

10.2. Give an example of a reducible subgroup **G** of $GL(2)$ for which the only elements of $GL(2)$ centralizing **G** are scalar matrices. (This shows that the complete converse of Problem 10.1 is false.)

10.3. If G is a completely reducible subgroup of $GL(\mathcal{V})$, and the only elements of $GL(\mathcal{V})$ centralizing G are scalars, then G is irreducible.

10.4. The derived group of $TL(n)$ is contained in $STL(n)$, and $STL(n)$ is nilpotent of class $n - 1$.

If $\mathbf{S}$ is any set of $n \times n$ matrices over $\mathcal{C}$, we define $\mathcal{C}\mathbf{S}$ to be the set of all finite linear combinations $\alpha_1\mathbf{s}_1 + \cdots + \alpha_m\mathbf{s}_m$ ($\alpha_i \in \mathcal{C}$, $\mathbf{s}_i \in \mathbf{S}$). We define addition and multiplication by scalars in an obvious manner for elements in $\mathcal{C}\mathbf{S}$.

10.5. Show that $\mathcal{C}\mathbf{S}$ is a vector space over $\mathcal{C}$ of dimension at most n^2.

10.6. Let G be a subgroup of $GL(\mathcal{V})$, and suppose that $\mathcal{V} = \mathcal{W}_1 \dot{+} \cdots \dot{+} \mathcal{W}_k$, where the $\mathcal{W}_i$ are invariant subspaces for G. Then G is isomorphic to a subgroup of the direct product $G|\mathcal{W}_1 \times \cdots \times G|\mathcal{W}_k$.

10.7. Let G be an irreducible subgroup of $GL(\mathcal{V})$, and let H be a normal subgroup of G. If $\mathcal{W}$ is a minimal invariant subspace for H in the underlying vector space $\mathcal{V}$, then $\mathcal{V} = \mathcal{W}x_1 \dot{+} \cdots \dot{+} \mathcal{W}x_k$ for some elements $x_1 = 1, \ldots, x_k$ in G.

10.8. If G is an irreducible subgroup of $GL(\mathcal{V})$, then each normal subgroup of G is completely reducible into k isomorphic irreducible components, where k divides the degree n of G. Any subnormal subgroup of a completely reducible group in $GL(\mathcal{V})$ is completely reducible.

10.9. Let G be an irreducible subgroup of $GL(\mathcal{V})$. If A is a normal abelian subgroup of G, then A corresponds to a diagonal group over a suitable basis for $\mathcal{V}$.

Let G be a subgroup of $GL(\mathcal{V})$. A *root space* $\mathcal{R}$ for G is a subspace of $\mathcal{V}$ which is invariant under G such that $G|\mathcal{R}$ is a group of scalars and such that $\mathcal{R}$ is maximal with respect to this property. Equivalently, $\mathcal{R}$ is a maximal subspace of $\mathcal{V}$ with the property that, for each $x \in G$ and each vector $v \neq 0$ in $\mathcal{R}$, v is an eigenvector for x.

10.10. Let A be a completely reducible abelian group in $GL(\mathcal{V})$, and let $\mathcal{R}_1, \ldots, \mathcal{R}_k$ be the distinct root spaces for A. Then $\mathcal{V}$ is the direct sum of these subspaces.

10.11. Let G be an irreducible subgroup of $GL(\mathcal{V})$. Let A be a normal abelian subgroup of G, and let $\mathcal{R}_1, \ldots, \mathcal{R}_k$ be the root spaces for A. Then, for each $x \in G$, the mapping $\mathcal{R}_i \rightarrow \mathcal{R}_i x$ ($i = 1, \ldots, k$) is a permutation of the set of root spaces. Hence, $G/C(A;G)$ is isomorphic to a transitive permutation group of degree k, and so, if G has degree n, $|G:C(A;G)|$ divides $n!$.

10.12. Let G be an irreducible subgroup of $GL(\mathcal{V})$ of degree n, and let H be a subgroup of index h in G. If H has an irreducible component of degree n/h which corresponds to a monomial group for a suitable basis, then G corresponds to a monomial group for some basis for $\mathcal{V}$.

10.13. Let G be a finite group with the following property. For each chain of subgroups $G \supseteq H \supset N \supseteq 1$, where N is a proper normal subgroup of H, either (a) H/N is abelian, or (b) there exists a normal abelian subgroup A of H such that N is properly contained in A, and A/N is not in the center of H/N.

If G is a linear group, then G corresponds to a monomial group over a suitable basis.

10.14. The condition in Problem 10.13 on the finite group G is satisfied in each of the following cases: (a) G is nilpotent; (b) $G'' = 1$; or (c) G is a solvable group in which each Sylow subgroup is abelian. Thus, in any of these cases, the conclusion of Problem 10.13 follows.

10.15. If G is a finite irreducible p-group in $GL(\mathcal{V})$, then the degree n is a power of p.

10.16. Let G be an irreducible subgroup of $GL(\mathcal{V})$. If $Z(G)$ is finite, then $Z(G)$ is cyclic.

10.17. Let G be a finite p-group in $GL(\mathcal{V})$. If the degree $n \geqslant 2$, then G' is reducible. [*Hint:* Use Problem 6.39.]

10.18. If G is a finite p-group in $GL(\mathcal{V})$ of degree n, then G is solvable of length l where $p^{l-1} \leqslant n$ (compare with Problem 6.44).

10.19. Let G be a subgroup of $GL(\mathcal{V})$. Put $C^* = \{\alpha 1 \mid \alpha \in \mathcal{C} \text{ and } \alpha \neq 0\} \subseteq GL(\mathcal{V})$, and $G_1 = GC^* \cap SL(\mathcal{V})$. Then $G \subseteq G_1C^*$, and G_1 is irreducible if and only if G is irreducible.

10.20. If G is an irreducible subgroup of $SL(\mathcal{V})$, then $Z(G)$ is a finite cyclic group whose order divides the degree n of G.

10.21. Let G be an irreducible group in $SL(\mathcal{V})$ of degree n, and let

$$1 = Z_0 \subseteq Z_1 \subseteq \cdots$$

be the upper central series of G. Then, for each $k \geqslant 0$, $x \in Z_{k+1}$ implies that $x^n \in Z_k$.

10.22. Let $\mathbf{G}$ be a subgroup of $D(n)$. If, for some integer $m > 0$, $\mathbf{x}^m = \mathbf{1}$ for all $\mathbf{x} \in \mathbf{G}$, then $| \mathbf{G} |$ divides m^n.

10.23. If G is a finite linear group of degree n, and, for some integer $m > 0$, $x^m = 1$ for all $x \in G$, then $| G |$ divides $n!m^n$.

10.24. Let G be an irreducible nilpotent subgroup of $SL(\mathcal{V})$ of degree n. Then G is finite. In particular, if G is nilpotent of class k, then $| G |$ divides both $n!n^{kn}$ and $n^{(k+1)n}$.

10.25. Let G be an irreducible nilpotent linear group of degree n. If G is of class k, then $| G:Z(G) |$ divides $n^{(k+1)n}$.

10.26. If G is a completely reducible nilpotent linear group of degree n, and G is of class k, then $| G:Z(G) | \leqslant n^{(k+1)n}$.

10.27. A completely reducible nilpotent linear group of degree n is solvable of length l where $2^{l-1} \leqslant n$ (compare with Problem 10.18).

10.28. Let $\mathbf{H}$ be a subgroup of a group $\mathbf{G} \subseteq GL(n)$. If the set

$$\{[\mathbf{x},\mathbf{y}] \mid \mathbf{x} \in \mathbf{G} \text{ and } \mathbf{y} \in \mathbf{H}\}$$

is a set of order m, then $| \mathbf{G}:\mathbf{C}(\mathbf{H};\mathbf{G}) | \leqslant m^{n^2}$.

10.29. Let G be an irreducible solvable subgroup of $SL(\mathcal{V})$ of degree n. Then G has a normal nilpotent subgroup N which properly contains the center Z of G such that $| N | \leqslant n^{4n}$.

10.30. Let G be a solvable linear group of degree n. Then G contains a normal reducible subgroup H whose index is at most n^{4n^3} in G.

10.31. If G is a solvable subgroup of $GL(\mathcal{V})$ of degree n, then G contains a subgroup M which corresponds to a triangular group over a suitable basis for $\mathcal{V}$ such that $| G:M | \leqslant n^{4n^4}$.

10.32. Let G be a solvable linear group of degree n. Then the solvable length l of G is bounded above by some number depending only on n.

10.33. Let G be an irreducible nilpotent linear group of degree n. Then G contains a normal abelian subgroup A whose index in G is at most 2^{n-1}. [*Hint:* Use Problem 6.46.]

10.34. Let G be a nilpotent linear group of degree n. Then G has a subgroup H which corresponds to a triangular group over a suitable basis such that $|G:H| \leqslant 2^{n-1}$ (compare with Problem 10.31).

10.35. Denote by $\mathbf{H}_n$ the generalized quaternion group of order 2^{n+1} defined in Problem 5.29. Then

$$\mathbf{H} = \bigcup_{n=2}^{\infty} \mathbf{H}_n$$

is a subgroup of $GL(2)$. Show that (a) $\mathbf{H}_n$ is nilpotent of class n; (b) $\mathbf{H}$ is a group in which every proper subgroup is nilpotent; and (c) $\mathbf{H}$ is not nilpotent, but its lower central series has the form: $\mathbf{H} = {}^0\mathbf{H} \supset {}^1\mathbf{H} = {}^2\mathbf{H} = \cdots$ (compare with Problem 10.32).

***10.36.** Let G be a subgroup of $GL(\mathcal{V})$, and let H be a subgroup of finite index in G. Then H is completely reducible if and only if G is completely reducible. (This generalizes 10.T.1.)

***10.37.** Let G be a solvable subgroup of $GL(\mathcal{V})$. Then G is completely reducible if and only if, for each $x \in G$, there is a basis for $\mathcal{V}$ over which x corresponds to a diagonal matrix. [*Hint:* Note that x corresponds to a diagonal matrix over a suitable basis for $\mathcal{V}$ if and only if the minimal polynomial for x has distinct roots.]

***10.38.** Let $\mathbf{G}$ be a finite subgroup of $GL(n)$ in which all entries of each matrix $\mathbf{x} \in \mathbf{G}$ are rational numbers. Then $\mathbf{G}$ is similar to a group of matrices with integral entries. [*Hint:* First try the case $n = 2$.]

***10.39.** Each element of $SL(n)$ is a commutator of elements in $SL(n)$.

***10.40.** Let $\mathbf{u}$ and $\mathbf{v}$ be elements of $SL(2)$ which have all their entries integers. If $\mathbf{uv} = \mathbf{vu}$, then $\langle \mathbf{u},\mathbf{v} \rangle$ is either cyclic or the direct product of a cyclic group by a group of order 2.

11 Representations and Characters

Let $\mathcal{C}$ be the field of complex numbers. A (matrix) *representation* of a group G is a homomorphism $\mathbf{R}$ of G into $GL(n)$ for some $n \geqslant 1$. We say that n is the *degree* of $\mathbf{R}$, and that $\mathbf{R}$ is *faithful* if its kernel is 1. The *character* corresponding to the representation $\mathbf{R}$ of G is the function χ of G into $\mathcal{C}$ defined by $\chi(x) = \operatorname{tr} \mathbf{R}(x)$. Two representations $\mathbf{R}$ and $\mathbf{S}$ of a group G are called *equivalent* if they have the same degree, say n, and if there exists a fixed $\mathbf{c} \in GL(n)$ such that $\mathbf{R}(x) = \mathbf{c}^{-1}\mathbf{S}(x)\mathbf{c}$ for all $x \in G$. A representation $\mathbf{R}$ of G is called *reducible* (*irreducible* or *completely reducible*) when the image $\mathbf{R}(G)$ is a reducible (irreducible or completely reducible, respectively) matrix group. In particular, every group G has an irreducible representation of degree 1 of the form $x \to 1$ ($x \in G$). The character of this representation is called the *identity* (or principal) *character* of G and is denoted by 1_G. Thus $1_G(x) = 1$ for all $x \in G$.

If G is a finite group of order g, and ϕ and ψ are functions of G into $\mathcal{C}$, then we define the *inner product*

$$\langle \phi, \psi \rangle = \frac{1}{g} \sum_{x \in G} \phi(x) \overline{\psi(x)},$$

where the bar denotes complex conjugate.

The following results are assumed. For their proofs see Hall [1], Sections 16.5 to 16.6 and Theorem 16.8.4; Schenkman [4], Sections 8.1 to 8.3; or Scott [5], Sections 12.1 to 12.2. Alternatively, see Curtis and Reiner [16] or Burrows [17].

11.T.1. Two representations of a finite group G have the same character if and only if they are equivalent. Thus, without ambiguity, we can call a character faithful, irreducible, etc., when that property is possessed by a corresponding representation. Similarly, we may define the kernel of a character to be the kernel of a corresponding representation.

11.T.2. A character χ of a group G is a *class function;* that is, if x and y are in the same conjugacy class of G, then $\chi(x) = \chi(y)$. If G is finite with conjugacy classes $C_1, \ldots, C_k$, then we shall write $\chi(x) = \chi_i$ for $x \in C_i$.

11.T.3. If $\mathbf{R}$ and $\mathbf{S}$ are irreducible representations of a group G and there is a matrix $\mathbf{c}$, of suitable dimensions, such that $\mathbf{cR}(x) = \mathbf{S}(x)\mathbf{c}$ for all $x \in G$, then either $\mathbf{c} = \mathbf{0}$ or $\mathbf{c}$ is nonsingular. In the latter case $\mathbf{R}$ is equivalent to $\mathbf{S}$. (This is known as *Schur's Lemma.*)

11.T.4. A finite group G with k conjugacy classes has exactly k distinct irreducible characters.

11.T.5. Let G be a finite group with k conjugacy classes, and let χ^a $(a = 1, 2, \ldots, k)$ be the distinct irreducible characters of G. Then we have the *character relations:*

$$\sum_{i=1}^{k} h_i \chi_i^a \overline{\chi_i^b} = g\delta_{ab} \qquad \text{for } a,b = 1, 2, \ldots, k,$$

and

$$\sum_{a=1}^{k} h_i \chi_i^a \overline{\chi_j^a} = g\delta_{ij} \qquad \text{for } i,j = 1, 2, \ldots, k,$$

where $g = |G|$, h_i is the order of the conjugacy class C_i of G, and the *Kronecker delta* $\delta_{st} = 1$ if $s = t$ and is 0 otherwise.

The first relations are equivalent to $\langle \chi^a, \chi^b \rangle = \delta_{ab}$. All together, the relations may be expressed as follows. The $k \times k$ matrix $\mathbf{u} = [u_{ij}]$, with $u_{ij} = \sqrt{h_j/g}\, \chi_j^i$, is a unitary matrix; that is, $\mathbf{uu}^* = \mathbf{u}^*\mathbf{u} = \mathbf{1}$, where $\mathbf{u}^*$ is the complex conjugate transpose of $\mathbf{u}$.

11.T.6. If G is a finite group of order g, and its irreducible characters χ^a $(a = 1, 2, \ldots, k)$ have degrees d_a $(a = 1, 2, \ldots, k)$, respectively, then $g = \sum_{a=1}^{k} d_a^2$ and $d_a \mid g$ $(a = 1, 2, \ldots, k)$. [*Note:* The degree $d_a = \chi^a(1)$.]

11.T.7. Let G be a finite group with irreducible characters

$$\chi^a \ (a = 1, 2, \ldots, k).$$

Then any character χ for G may be written $\chi = \sum_{a=1}^{k} c_a\chi^a$, where the c_a are nonnegative integers. This representation of χ is unique and, in fact, $c_a = \langle\chi,\chi^a\rangle$. We call c_a the *multiplicity* of χ^a in χ, and χ^a is an (irreducible) *constituent* of χ when $c_a > 0$. Finally, $\langle\chi,\chi\rangle = \sum_{a=1}^{k} c_a^2$, and so χ is irreducible if and only if $\langle\chi,\chi\rangle = 1$.

11.T.8. Let **R** be a representation of degree n of a finite group G. Then, for each $x \in G$, $\mathbf{R}(x)$ is a matrix which is similar to a diagonal matrix whose diagonal entries are roots of unity. Moreover, if χ is the character corresponding to **R**, then $\chi(x^{-1}) = \overline{\chi(x)}$ (complex conjugate).

PROBLEMS

11.1. Show that the symmetric group S_3 has exactly three irreducible characters, and calculate these characters.

11.2. Calculate the irreducible characters of the alternating group A_4.

11.3. Let G be a finite abelian group of the form: $G = C_1 \times C_2 \times \cdots \times C_m$, where C_i is a cyclic group of order h_i $(i = 1, 2, \ldots, m)$. Show that all irreducible characters of G have degree 1, and calculate these characters.

11.4. The infinite dihedral group G defined in Problem 5.11 has an infinite (uncountable) number of inequivalent representations of degree 2, all of which are both faithful and irreducible.

11.5. There are two nonisomorphic groups of order 8 which have the same character table $[\chi_j^i]$. [*Hint:* See Problem 3.3.]

11.6. Let $\mathbf{w}$ be a nonsingular matrix of degree n. Let u and v be permutations in S_n. Define $\mathbf{w}_1$ as the matrix whose ith row is the same as the i^uth row of $\mathbf{w}$, and define $\mathbf{w}_2$ as the matrix whose jth column is the j^vth column of $\mathbf{w}$ (for $i,j = 1, 2, \ldots, n$). If $\mathbf{w}_1 = \mathbf{w}_2$, then u and v leave fixed the same number of letters.

Let G be a finite group. For any representation $\mathbf{R}$ of G, we have the *conjugate* representation $\mathbf{R}^*$ of G defined by $\mathbf{R}^*(x) = \overline{\mathbf{R}(x)}$ $(x \in G)$; that is, the entries in the matrix $\mathbf{R}^*(x)$ are the complex conjugates of the corresponding entries of $\mathbf{R}(x)$. The character χ of $\mathbf{R}$ is called *real* if it equals its conjugate character χ^* (corresponding to $\mathbf{R}^*$). We also define the *inverse* of a conjugacy class C_i of G to be the conjugacy class

$$C_{i^*} = \{x^{-1} \mid x \in C_i\},$$

and call C_i *self-inverse* if it equals its inverse class.

11.7. Let G be a finite group. Then the number of real irreducible characters of G is equal to the number of self-inverse conjugacy classes of G.

11.8. If G is a finite group of odd order, then the only real irreducible character of G is the identity character 1_G.

11.9. If G is a group of odd order n, and G has k conjugacy classes, then $k \equiv n \pmod{16}$.

11.10. A finite p-group has a faithful irreducible representation if and only if its center $Z(G)$ is cyclic.

11.11. If a finite group G has a faithful representation $\mathbf{R}$ of degree n, and n is less than the smallest prime dividing $|G|$, then G is abelian.

11.12. Let G be a simple group of order n, and let p be a prime dividing n. If G has more than n/p^2 conjugacy classes, then the Sylow p-groups of G are abelian.

Let H be a subgroup of a group G. If $\mathbf{R}$ is a representation of G, then the restriction $\mathbf{R}|H$ is a representation of H defined by $(\mathbf{R}|H)(x) = \mathbf{R}(x)$ $(x \in H)$. The character $\chi|H$ of $\mathbf{R}|H$ is the restriction to H of the character χ corresponding to $\mathbf{R}$.

Conversely, we may obtain characters of G from characters of a subgroup H. Let us suppose that H is a subgroup of finite index in G, and that $r_1, r_2, \ldots, r_n$ is a set of left coset representatives of H in G. Let $\mathbf{R}$ be a representation of degree m for H, and define $\mathbf{R}^G$ (the *induced representation*) on G by defining

$$\mathbf{R}^G(x) = [\breve{\mathbf{R}}(r_i^{-1}xr_j)] \qquad (x \in G)$$

as a matrix of degree mn, where the $m \times m$ blocks $\breve{\mathbf{R}}(r_i^{-1}xr_j)$ are defined by

$$\breve{\mathbf{R}}(y) = \begin{cases} \mathbf{R}(y) & \text{if } y \in H, \\ \mathbf{0} & \text{otherwise.} \end{cases}$$

Precisely as in the case for monomial representations (see Chapter 7), it can be shown that $\mathbf{R}^G$ is a representation of degree mn for G. The *induced character* χ^G corresponding to $\mathbf{R}^G$ is given in terms of the character χ corresponding to $\mathbf{R}$ by

$$\chi^G(x) = \sum_{i=1}^{k} \breve{\chi}(r_i^{-1}x_ir) \qquad (x \in G),$$

where $\breve{\chi}(y) = \chi(y)$ if $y \in H$ and is 0 otherwise.

11.13. Let H be a subgroup of a finite group G. If χ is a character of H, then

$$\chi^G(x) = \frac{1}{|H|}\sum_{y\in G} \breve{\chi}(y^{-1}xy) \qquad (x \in G).$$

In particular, χ^G does not depend on the choice of coset representatives for H in G.

11.14. Let H be a subgroup of a finite group G, and let χ and ψ be characters of G and H, respectively. Then $\langle\psi,\chi|H\rangle = \langle\psi^G,\chi\rangle$. [*Note:* The first inner product is over H, but the second is over G.] The following result then follows from 11.T.7.

Frobenius Reciprocity Theorem: If H is a subgroup of a finite group G, and χ and ψ are irreducible characters of G and H, respectively, then the multiplicity with which χ occurs as a constituent of ψ^G is equal to the multiplicity with which ψ occurs as a constituent of $\chi|H$.

11.15. Let G be the dihedral group D_7 defined in Problem 1.6, and let H be the subgroup $\langle b\rangle$. Calculate the characters of G which are induced from the irreducible characters of H, and express them in terms of irreducible characters of G. Hence, calculate all irreducible characters of G.

11.16. What are the characters of the alternating group A_5 which are induced from irreducible characters of the subgroup A_4 (see Problem 11.2)? Use them to calculate all irreducible characters of A_5.

11.17. Let H be a subgroup of index m in a group G. For any irreducible character χ of G, there exists an irreducible character ψ of H such that $m\psi(1) \geqslant \chi(1)$. In particular, if H is abelian, then every irreducible character χ of G has degree at most m.

Any permutation group G of degree n has a linear representation of degree n in terms of permutation matrices. This is defined by $x \to [\xi_{ij}]$, where $\xi_{ij} = 1$ if $j = i^x$ and is 0 otherwise. We call the character π of this representation the *permutation character* of G, and note that $\pi(x)$ is a nonnegative integer equal to the number of letters left fixed by x, for each $x \in G$.

11.18. Let G be a permutation group of degree n with the permutation character π. If G has s orbits, then there exist s subgroups $H_1, H_2, \ldots, H_s$ of G such that $\pi = (1_{H_1})^G + \cdots (1_{H_s})^G$. In particular, if G is transitive, then π is the character induced from the identity character 1_H of a stabilizer H of G.

11.19. Let G be a permutation group of degree n with permutation character π. Then G has s orbits, where $s = \langle \pi, 1_G \rangle$. Furthermore, if G is transitive, and H is a stabilizer of G, then H has t orbits, where $t = \langle \pi, \pi \rangle$.

11.20. Let G be a transitive permutation group of degree n with permutation character π. Then, for any irreducible character χ of $G, \langle \pi, \chi \rangle \leqslant \chi(1)$. There is equality if and only if the kernel of χ contains all the stabilizers of G.

11.21. Let N be a normal subgroup of a finite group G. If $\alpha^1, \alpha^2, \ldots, \alpha^s$ are all the irreducible characters of G such that $N \subseteq \ker \alpha^i$, then the group G/N has exactly s irreducible characters $\beta^1, \beta^2, \ldots, \beta^s$. These are defined by $\beta^i(Nx) = \alpha^i(x)$ for each $Nx \in G/N$ and $i = 1, 2, \ldots, s$.

11.22. Let N be a normal subgroup of a finite group G, and let χ be an irreducible character of G such that the kernel of χ does not contain N. If x in G does not commute with any nontrivial element of N, then $\chi(x) = 0$.

Let $\mathbf{x} = [\xi_{ij}]$ and $\mathbf{y} = [\eta_{ij}]$ be two square matrices of degrees m and n, respectively. We define the *tensor* (or Kronecker) product $\mathbf{x} \otimes \mathbf{y}$ as the matrix of degree mn of the form

$$\begin{bmatrix} \mathbf{x}\eta_{11} & \mathbf{x}\eta_{12} & \cdots & \mathbf{x}\eta_{1n} \\ \mathbf{x}\eta_{21} & \mathbf{x}\eta_{22} & \cdots & \mathbf{x}\eta_{2n} \\ \cdot & & & \cdot \\ \cdot & & & \cdot \\ \cdot & & & \cdot \\ \mathbf{x}\eta_{n1} & \mathbf{x}\eta_{n2} & \cdots & \mathbf{x}\eta_{nn} \end{bmatrix} \tag{11.1}$$

where the $m \times m$ blocks $\mathbf{x}\eta_{ij}$ have the form

$$\begin{bmatrix} \xi_{11}\eta_{ij} & \xi_{12}\eta_{ij} & \cdots & \xi_{1m}\eta_{ij} \\ \xi_{21}\eta_{ij} & \xi_{22}\eta_{ij} & \cdots & \xi_{2m}\eta_{ij} \\ \cdot & & & \cdot \\ \cdot & & & \cdot \\ \cdot & & & \cdot \\ \xi_{m1}\eta_{ij} & \xi_{m2}\eta_{ij} & \cdots & \xi_{mm}\eta_{ij} \end{bmatrix}.$$

If $\mathbf{x}_1$ and $\mathbf{y}_1$ are also two matrices of degrees m and n, respectively, then straightforward computation shows that

$$(\mathbf{x} \otimes \mathbf{y})(\mathbf{x}_1 \otimes \mathbf{y}_1) = (\mathbf{x}\mathbf{x}_1) \otimes (\mathbf{y}\mathbf{y}_1). \tag{11.2}$$

It follows immediately that, if $\mathbf{G}$ and $\mathbf{H}$ are matrix groups of degrees m and n, respectively, then $\mathbf{G} \otimes \mathbf{H} = \{\mathbf{x} \otimes \mathbf{y} \mid \mathbf{x} \in \mathbf{G} \text{ and } \mathbf{y} \in \mathbf{H}\}$ is a matrix group of degree mn which is isomorphic to the (external) direct product $\mathbf{G} \times \mathbf{H}$.

11.23. Let $\mathbf{x}$ and $\mathbf{y}$ be matrices of degrees m and n, respectively. Then

(a) $\operatorname{tr}(\mathbf{x} \otimes \mathbf{y}) = (\operatorname{tr} \mathbf{x})(\operatorname{tr} \mathbf{y})$;
(b) $\det(\mathbf{x} \otimes \mathbf{y}) = (\det \mathbf{x})^n(\det \mathbf{y})^m$.

11.24. If χ and ψ are characters for a group G, then the function $\chi\psi$ of G into $\mathfrak{C}$ defined by $\chi\psi(x) = \chi(x)\psi(x)$ $(x \in G)$ is also a character for G.

11.25. Let $\mathbf{R}$ be an irreducible representation of degree n of a group which is the direct product of two normal subgroups G and H. Then $\mathbf{R}$ is equivalent to a representation $\mathbf{R}_1$ such that $\mathbf{R}_1(G \times H) = \mathbf{S}(G) \otimes \mathbf{T}(H)$, where $\mathbf{S}$ and $\mathbf{T}$ are irreducible representations of G and H, respectively. Thus, for any irreducible character χ of $G \times H$, there exist irreducible characters α and β of G and H, respectively, such that $\chi(xy) = \alpha(x)\beta(y)$ $(x \in G; y \in H)$. [*Hint:* Use Problem 10.7.]

11.26. Let χ be an irreducible character of degree d for a finite group G, and let K be the kernel of χ. Then (a) $|\chi(x)| = d$ if and only if $xK \in Z(G/K)$, and (b) $\chi(x) = d$ if and only if $x \in K$.

11.27. Suppose that χ is a faithful character of degree d for a finite group G of order g. We define the powers of χ by using Problem 11.24:

$$\chi^{(0)} = 1_G \quad \text{and} \quad \chi^{(s)} = \chi^{(s-1)}\chi \quad \text{for } s = 1, 2, \ldots$$

If $\chi(x)$ $(x \in G)$ takes exactly r different values, say $\alpha_1, \alpha_2, \ldots, \alpha_r$, then each irreducible character ψ of G occurs as a constituent of at least one of the characters $\chi^{(s)}$ $(s = 0, 1, \ldots, r - 1)$.

In particular, the sum of the degrees of the irreducible characters of G is at most $(d^r - 1)/(d - 1)$.

11.28. Let χ be a character for a finite group G, and let ψ be a character for some subgroup H of G. Then $\{(\chi|H)\psi\}^G = \chi\psi^G$.

11.29. Let N be a normal subgroup of index n in a finite group G. If χ is a character of G such that $\phi = \chi|N$ is an irreducible character, then $\langle\phi^G,\phi^G\rangle = n$.

11.30. Let χ be an irreducible character of a finite group G of order g. Let N be a normal subgroup of index n in G, and suppose that $\chi|N$ is an irreducible character for N. Then, for each irreducible character ψ of G whose kernel contains N, $\chi\psi$ is an irreducible character for G.

11.31. Let G be a finite group with a normal series

$$G = G_0 \supseteq G_1 \supseteq G_2 \supseteq \cdots \supseteq G_n = 1.$$

If m of the factor groups G_{i-1}/G_i ($i = 1, 2, \ldots, n$) are nonabelian, then G has an irreducible character of degree $\geqslant 2^m$.

11.32. Let G be a finite group of order g with a faithful irreducible representation $\mathbf{R}$ of degree n. If $G' \subseteq Z(G)$, then $|G:Z(G)| = n^2$. [*Hint:* First show that the character corresponding to $\mathbf{R}$ is zero outside $Z(G)$.]

Let $\mathcal{L}$ be the set of all $n \times n$ matrices over $\mathcal{C}$. Then $\mathcal{L}$ is a vector space of dimension n^2 over $\mathcal{C}$ (see Problem 10.5). Let $\mathbf{G}$ be a subgroup of $GL(n)$. Then we can define a homomorphism R of $\mathbf{G}$ into the group $GL(\mathcal{L})$ by

$$R(\mathbf{x}): \mathbf{u} \to \mathbf{ux} \qquad (\mathbf{u} \in \mathcal{L}), \tag{11.3}$$

for each $\mathbf{x} \in \mathbf{G}$.

11.33. Let $\mathbf{G}$ be an irreducible subgroup of $GL(n)$, and let R be the mapping defined by (11.3). Then any subspace $\mathfrak{M} \neq \mathbf{0}$ of $\mathcal{L}$, which is minimal with respect to the property $\mathfrak{M}\mathbf{x} \subseteq \mathfrak{M}$ for all $\mathbf{x} \in \mathbf{G}$, has dimension n.

11.34. Let $\mathbf{G}$ be a subgroup of $GL(n)$, and define

$$\mathbf{G}^\perp = \{\mathbf{u} \in \mathcal{L} \mid \operatorname{tr}(\mathbf{ux}) = 0 \qquad \text{for all } \mathbf{x} \in \mathbf{G}\}.$$

If $\mathbf{G}$ is irreducible, then $\mathbf{G}^\perp = \mathbf{0}$. It follows that $\mathcal{C}\mathbf{G}$ has dimension n^2; that is, $\mathcal{C}\mathbf{G} = \mathcal{L}$ (see Problem 10.5).

11.35. Let $\mathbf{G}$ be an irreducible subgroup of $GL(n)$. If $\operatorname{tr}\mathbf{x}$ ($\mathbf{x} \in \mathbf{G}$) takes only m different values, then $\mathbf{G}$ is finite and $|\mathbf{G}| \leqslant m^{n^2}$.

11.36. Let G be a group with only a finite number of conjugacy classes. If G has a faithful matrix representation, then G is finite.

11.37. Let $\mathbf{G}$ be a subgroup of $GL(n)$. If, for some integer $m > 0$, $\mathbf{x}^m = \mathbf{1}$ for all $\mathbf{x} \in \mathbf{G}$, then $\mathbf{G}$ is finite and $|\mathbf{G}| \leqslant m^{n^3}$ (compare with Problem 10.23).

11.38. Let $\mathbf{G}$ be a subgroup of $GL(n)$ in which each $\mathbf{x} \in \mathbf{G}$ is similar to a matrix in $STL(n)$; that is, every eigenvalue of $\mathbf{x}$ is 1. Then $\mathbf{G}$ is conjugate in $GL(n)$ to a subgroup of $STL(n)$.

11.39. Let χ be an irreducible character of a finite group G. Then $\chi(1)^2 \leqslant |G:Z(G)|$ (compare with Problems 11.17 and 11.32).

***11.40.** Let G be a finite group for which each matrix representation is equivalent to a representation in terms of monomial matrices. Then G is solvable (compare with Problem 10.13).

Solutions

Solutions for Chapter 1—Subgroups

1.1. G has order 6 and its subgroups are: 1, $\{1,x,x^2\}$, $\{1,y\}$, $\{1,xy\}$, $\{1,x^2y\}$, and G.

1.2. G has order 12 and has a normal subgroup $\{1,y,x^{-1}yx,xyx^{-1}\}$ (see [18], page 465).

1.3. From the first relation, $x^2y^4x^{-2} = xy^6x^{-1} = y^9$. Using the second relation, we have $y^9 = y^{-1}x^2y \cdot y^4 \cdot y^{-1}x^{-2}y = x^3y^4x^{-3}$, and so $xy^4 = y^4x$. Hence we find $x = y = 1$, and $G = 1$.

1.6. (c) The conjugacy classes of D_k are: $\{1\}$, $\{a, ab, \ldots, ab^{k-1}\}$, and $\{b^i,b^{-i}\}$, for $1 \leqslant i \leqslant \frac{1}{2}(k-1)$, if k is odd; and $\{1\}$, $\{a, ab^2, \ldots, ab^{k-2}\}$, $\{ab, ab^3, \ldots, ab^{k-1}\}$, $\{b^{k/2}\}$, and $\{b^i,b^{-i}\}$, for $1 \leqslant i \leqslant \frac{1}{2}(k-2)$, if k is even.

1.7. $H \subseteq N(H;G)$, and so by 1.T.3, H has a finite number of conjugates $H = H_1, H_2, \ldots, H_m$, say, in G. Since $|G:H_i| = |G:H|$ for each i, we can apply 1.T.2 to show successively that $H_1, H_1 \cap H_2, \ldots,$ $H_1 \cap H_2 \cap \cdots \cap H_m$ are each of finite index in G. The last of these subgroups is the required normal subgroup of G.

1.8. By 1.T.1, $| G{:}A \cap B |$ is divisible by both $| G{:}A |$ and $| G{:}B |$, and so by their least common multiple. Since $| G{:}A |$ and $| G{:}B |$ are relatively prime, their least common multiple is $| G{:}A | \, | G{:}B |$, and so this is at most $| G{:}A \cap B |$. The result now follows from 1.T.2.

1.9. Let $| G{:}H | = h$ and $| G | = n$. Since $H \subseteq N(H;G)$, therefore, by 1.T.3, H has at most h conjugates in G. Since all subgroups have the identity element in common, the number of distinct elements in these conjugates of H is at most $1 + (| H | - 1)h = n - h + 1 < n$.

1.10. By Problem 1.7 we can find a normal subgroup N of G contained in H with G/N a finite group. By Problem 1.9 we can choose a coset Nx in G/N such that Nx is not contained in any conjugate $y^{-1}Hy/N$ of H/N in G/N. Then $x \notin y^{-1}Hy$ for any $y \in G$.

1.11. Let M be a maximal subgroup of G. Since all maximal subgroups of G are conjugate to M, we can use Problem 1.10 to find an element x in G which does not lie in any maximal subgroup. Then $\langle x \rangle$ cannot be a proper subgroup of G, and so $\langle x \rangle = G$ (see [19]).

1.12. It is sufficient to show that if $u \in \langle S \rangle$ can be written as a product $s_1s_2 \cdots s_m$ $(s_i \in S)$, with $m > (n - 1) \, | S |$, then it may be written as a product of fewer elements in S.

Since $m > (n - 1) | S |$, one element s in S must appear at least n times in the product. Then $u = s^n s'_{n+1} s'_{n+2} \cdots s'_m$, where the elements s'_i are some elements conjugate (under powers of s) to the elements in $s_1s_2 \cdots s_m$ which are not collected in the factor s^n. Since $s^n = 1$ (by hypothesis), u is a product of $m - n$ elements of S (see [2], Section 53).

1.13. Let $N = N(H;K)$. Then $\langle H,N \rangle = HN$ because H is normal in this group (1.T.5). Moreover, H is a Hall subgroup of HN, by 1.T.1, and so $| HN{:}H | = | N{:}H \cap N |$ is relatively prime to $| H | = | K |$. Since N is a subgroup of K, this implies that $| N{:}H \cap N | = 1$; that is, $N \subseteq H$. Since $H \cap K \subseteq N$, the result follows.

1.14. If N is any subset of G, and x and y in G both centralize N, then x^{-1} and xy also centralize N. Thus $C(N;G)$ is a subgroup. If N is normal in G, and $x \in C(N;G)$, then, for all $y \in G$ and $u \in N$, $yuy^{-1} \in N$. Therefore

$x^{-1}yuy^{-1}x = yuy^{-1}$; that is, $u(y^{-1}xy) = (y^{-1}xy)u$. Thus $y^{-1}xy \in C(N;G)$ as required.

If N is a characteristic subset of G, and $x \in C(N;G)$, then, for any automorphism α of G and all $u \in N$, $u^{\alpha^{-1}} \in N$. Therefore $x^{-1}u^{\alpha^{-1}}x = u^{\alpha^{-1}}$. Under the automorphism α, this equation gives $(x^\alpha)^{-1}ux^\alpha = u$, and so $x^\alpha \in C(N;G)$ as required.

1.15. It is clear that N is a normal subgroup of G. If G/N is a cyclic group, let Nx be a generator of this group. Then $G = \langle x,N\rangle$, and so G is generated by a set in which each pair of elements commute. Hence G is abelian.

1.16. If $x,y \in G$, then $(xy)^2 = 1 = x^2 = y^2$. Thus $x(yx)y = x(xy)y$, and so $xy = yx$.

1.17. We note that $|G:N|$ is divisible by $|HN:N| = |H:H \cap N|$. Since $|G:N|$ and $|H|$ are relatively prime, $|H:H \cap N|$, which divides both, equals 1. Hence $H \cap N = H$; that is, $H \subseteq N$.

1.18. Since $|HN:H| = |N:H \cap N|$ divides both of the relatively prime numbers $|N|$ and $|G:H|$, therefore $|N:H \cap N| = 1$. Thus $N \subseteq H$.

1.19. The elements of $G/N \times G/M$ are of the form (Nx,My) $(x,y \in G)$. The mapping $\theta: x \to (Nx,Mx)$ $(x \in G)$ is readily verified to be a homomorphism of G into $G/N \times G/M$, and its kernel is $\{x \in G \mid Nx = N$ and $Mx = M\} = M \cap N$. The result follows.

1.20. Define $q_i = n/p_i^{k_i}$ $(i = 1, 2, \ldots, s)$. Since the greatest common divisor of all the q_i is 1, we can find integers m_i such that $q_1m_1 + q_2m_2 + \cdots + q_sm_s = 1$. Write $x_i = x^{q_im_i}$. Then x_i has order $p_i^{k_i}$ for $i = 1, 2, \ldots, s$, and $x = x_1x_2 \cdots x_s$.

To prove uniqueness we proceed as follows. By a simple induction argument we see that the order of the group $\langle y_2,y_3, \ldots, y_s\rangle = \langle y_2\rangle\langle y_3\rangle \cdots \langle y_s\rangle$ is relatively prime to p_1, and so the intersection of this group with $\langle y_1\rangle$ is 1. Since $1 = x^n = y_1^n y_2^n \cdots y_s^n$, we conclude that $y_1^n = 1$. Similarly, we see that the order of y_i divides n for $i = 1, 2, \ldots, s$. It is then easily seen that $y_j^{q_im_i}$ equals 1 if $i \neq j$ and equals y_i if $i = j$. Since $x = y_1y_2 \cdots y_s$, we find $x_i = x^{q_im_i} = y_i$ $(i = 1, 2, \cdots, s)$.

1.21. By 1.T.3, $|H:N(X;H)|$ is finite because $X = \langle x\rangle$ has only a finite number of conjugates in H. But clearly $C(X;H) \subseteq H \cap x^{-1}Hx$, so it remains to show that $C(X;H)$ has finite index in $N(X;H)$. Indeed we show it has index at most 2 if X is infinite, and at most h if $|X| = h$ is finite. For suppose that $y \in N(X;H)$. Then $\langle y^{-1}xy \rangle = y^{-1}Xy = X$, so $y^{-1}xy$ generates X. Moreover, if $y_1, \ldots, y_r$ are representatives of distinct right cosets of $C(X;H)$ in $N(X;H)$, then $y_iy_j^{-1} \notin C(X;H)$ for any $i \neq j$; hence $y_i^{-1}xy_i$ $(i = 1, \ldots, r)$ are distinct generators of X. But X has two generators (x and x^{-1}) when X is infinite, and at most h generators when $|X| = h$. Therefore $|N(X;H):C(X;H)|$ is at most 2 or h in the respective cases.

1.22. Let $H = H_1, H_2, \ldots, H_s$ be the subgroups conjugate to H in G. Using Problem 1.21 we find, by induction on k, that $\bigcap_{i=1}^{k} H_i$ is of finite index in H for $k = 1, 2, \ldots, s$. The last of these subgroups is normal in G (see [20] and [21]).

1.23. Let x and y in F have m and n conjugates, respectively, in G. For each $u \in G$, $u^{-1}x^{-1}u = (u^{-1}xu)^{-1}$ and $u^{-1}(xy)u = (u^{-1}xu)(u^{-1}yu)$. Thus x^{-1} and xy have at most m and mn conjugates in G, respectively. Therefore F is a subgroup of G. Similarly, for each automorphism α of G, x^α has the same number of conjugates as x. Therefore F is a characteristic subgroup (see [22]).

1.24. By the definition of H, $u^{-1}xu = 1^{-1}x1 = x$ for $u \in H$, and $x \notin H$ implies that $u = 1$. Therefore $H \cap C(x;G) = 1$. On the other hand, for all $y \in G$, we can find $u \in H$ such that $y^{-1}xy = u^{-1}xu$. Thus $y \in C(x;G)u \subseteq C(x;G)H$. Therefore $G = C(x;G)H$.

1.25. Let $v \in H$, $y \in G$, and put $x = yvy^{-1}$. We wish to show that $x \in H$ for all such v and y. Let us suppose that $x \notin H$. Then, since H is special, $v = y^{-1}xy = u^{-1}xu$ for some $u \in H$. Hence $uvu^{-1} = x$ lies in H, which is the desired contradiction.

1.26. Let $x \in G$ and $x \notin H$. Because H is special, the elements $u^{-1}xux^{-1}$ $(u \in H)$ are all different, and so comprise a set of the same order as H. Since H is finite and the elements $u^{-1}(xux^{-1})$ all lie in H by Problem 1.25, $H = \{u^{-1}xux^{-1} \mid u \in H\}$. Thus $Hx = \{u^{-1}xu \mid u \in H\}$ which equals $\{y^{-1}xy \mid y \in G\}$ because H is special.

1.27. Putting x^{-1} for x, it follows as in the proof of Problem 1.26 that $H \supseteq \{u^{-1}x^{-1}ux \mid u \in H\} = \{y^{-1}x^{-1}yx \mid y \in G\}$ for each $x \in G$. There is equality when H is finite.

1.28. Let $y,z \in C(x;G)$. Then $y^{-1}z^{-1}yz$ lies in both $C(x;G)$ and H, by Problem 1.27. Therefore, by Problem 1.24, $y^{-1}z^{-1}yz = 1$; that is, $yz = zy$.

1.29. Put $V = U \cap y^{-1}Uy$ with $y \notin U$, and suppose that $V \neq 1$. Then there exists $v = y^{-1}uy \in V$ with $u \neq 1$ in U, and $u^{-1}v = u^{-1}y^{-1}uy \in H$, by Problem 1.27. Since $U \cap H = 1$ by Problem 1.24, $u^{-1}v = 1$; that is, $u = v$. Therefore $v \in C(x;G) \cap C(y;G)$, and, in particular, $x^{-1}y^{-1}xy \in C(v;G)$. But $H \cap C(v;G) = 1$ by Problem 1.24, and so Problem 1.27 shows that $x^{-1}y^{-1}xy = 1$. Thus $y \in C(x;G) = U$, contrary to hypothesis. (For the results of Problems 1.24 to 1.29 see [23].)

1.30. We define the product of two elements x and y in G as follows. Choose n such that x and y both lie in G_n. (This is possible because we are dealing with a chain.) Then define xy as in G_n. It is now straightforward to verify that this definition is independent of the particular choice of G_n, that G is a group under this product, and that each G_n is a subgroup of G. The uniqueness is obvious. Finally, if each G_n is abelian, then $xy = yx$ for all $x,y \in G$, from the definition above (see [2], Section 7).

1.31. Let $x \in H$ and $y \in G$. Then choose G_n so that $x \in H_n$ and $y \in G_n$. By hypothesis, $y^{-1}xy \in H_n \subseteq H$. Therefore $y^{-1}Hy \subseteq H$ for all $y \in G$.

1.32. Let $x_1, x_2, \ldots, x_n$ be a finite set of generators for G. Let $G_0, G_1, \ldots,$ be a proper ascending chain of subgroups of G with its union equal to G. Choose G_{m_i} so that $x_i \in G_{m_i}$, and let m be the maximum of $m_1, m_2, \ldots, m_n$. Then $G = \langle x_1, x_2, \ldots, x_n \rangle \subseteq G_m$; that is, the chain terminates at G_k for some $k \leqslant m$.

1.33. Let $\mathcal{S}$ be any nonempty set of subgroups of G. Let G_0 be any subgroup in $\mathcal{S}$, and choose successively (as long as possible) $G_1, G_2, \ldots,$ from $\mathcal{S}$ such that G_{i+1} properly contains G_i ($i = 1, 2, \ldots$). By hypothesis, this chain of subgroups is finite and terminates at, say, G_m. Then G_m is a subgroup in $\mathcal{S}$ which is not contained in any other subgroup in $\mathcal{S}$.

1.34. Let $\mathcal{S}$ be the set of subgroups appearing in the chain. By hypothesis, there is at least one of these subgroups not contained in any other. Because the chain is proper, it must terminate at this subgroup.

1.35. The solution is analogous to those of Problems 1.33 and 1.34.

1.36. If each subgroup of G is finitely generated, then, from Problems 1.32 and 1.33, G satisfies the maximal condition on subgroups.

Conversely, let us suppose that G satisfies the maximal condition on subgroups, and let H be a subgroup of G. We choose successively (as long as possible) elements $x_1, x_2, \ldots,$ of H with the property that the ascending chain of subgroups:

$$H_0 = 1, \qquad H_1 = \langle x_1 \rangle, \qquad H_2 = \langle x_1, x_2 \rangle, \ldots$$

is proper; that is, x_n is chosen so that $x_n \notin H_{n-1}$. By Problem 1.34, this chain is finite. This implies that, for some m, $H_m = H$; that is, H is generated by the elements $x_1, x_2, \ldots, x_m$.

1.37. We suppose that G satisfies the maximal condition on subgroups. Clearly, each subgroup does also. If G/N is a factor group, and $\mathcal{S}$ is a nonempty set of subgroups of G/N, then define $\mathcal{S}^* = \{H \subseteq G \mid N \subseteq H$ and $H/N \in \mathcal{S}\}$. By hypothesis, there exists $M \in \mathcal{S}^*$ which is not contained in any other subgroup in $\mathcal{S}^*$. Then M/N has the same property in $\mathcal{S}$. The case for the minimal condition is analogous.

1.38. Let $ac \in A(B \cap C)$, where $a \in A$ and $c \in B \cap C$. Then $ac \in AB$, and $ac \in aC = C$. Therefore $A(B \cap C) \subseteq AB \cap C$. On the other hand, if $ab \in AB \cap C$, where $a \in A$ and $b \in B$, then $b \in a^{-1}C = C$, and so $ab \in A(B \cap C)$. Thus $AB \cap C \subseteq A(B \cap C)$, and the result follows.

1.39. From Problem 1.38, $A = A(A \cap K) = A(B \cap K) = AK \cap B = BK \cap B = B$.

1.40. Let us suppose that both N and G/N satisfy the maximal condition on subgroups. Let $\mathcal{S}$ be a nonempty set of subgroups of G, and define $\mathcal{S}_1 = \{H \cap N \mid H \in \mathcal{S}\}$. We may choose $M_1 \in \mathcal{S}$ so that the subgroup $M_1 \cap N$ is not contained in any other subgroup in $\mathcal{S}_1$. Now define $\mathcal{S}_2 = \{HN/N \mid H \in \mathcal{S}$ and $H \cap N = M_1 \cap N\}$. Then we may choose $M \in \mathcal{S}$ such that MN/N is in $\mathcal{S}_2$, and MN/N is not contained in any other subgroup in $\mathcal{S}_2$. We now assert that M is not contained in any other subgroup in $\mathcal{S}$. Let us suppose that $L \in \mathcal{S}$ and $M \subseteq L$. Then $M \cap N = M_1 \cap N \subseteq L \cap N$, and so, by the choice of M_1, $M \cap N = L \cap N$. Similarly, $MN/N = LN/N$; that is, $MN = LN$, by the choice of M. Hence, by Problem 1.39, $M = L$ as required. The case of the minimal condition on subgroups is analogous. (Compare Problem 1.40 with the analogous result for rings given in [24] ,page 155.)

1.41. It is sufficient to show that, if M is a maximal element in the set of all normal subgroups N of G for which $G/N \simeq G$, then $M = 1$. Let ϕ be an isomorphism of G/M onto G. Then we define a mapping θ of G into itself as follows: if $(Mx)^\phi = y$ and $(My)^\phi = z$, then $x^\theta = z$. It is readily verified that θ is a homomorphism of G onto itself with the kernel $K = \{x \in G \mid (Mx)^\phi \in M\}$. In particular, $M \subseteq K$, and $G/K \simeq G$. Now let us suppose that $u \neq 1$ lies in M. Then, for some $v \in K$, $(Mv)^\phi = u$, and $v \notin M$ because $u \neq 1$. Therefore K properly contains M, contrary to the choice of M. Hence we conclude that $M = 1$, and the result follows (see [25], page 234).

1.42. Since $x^{-1}Hx \cap H = 1$ for all $x \notin H$, therefore $N(H;G) = H$. Thus H has $d = n/m$ conjugates in G. Since any two of these conjugates have only the identity element in common, their set union contains $d(m-1)+1$ elements. There are then $n - d(m-1) - 1 = (n/m) - 1$ elements remaining.

1.43. Since A is an abelian group, $A \cap B$ is normal in A. Similarly, $A \cap B$ is normal in B, and so it is normal in $\langle A,B\rangle$.

1.44. Let us suppose that G contains only the two normal subgroups 1 and G. Let M be a maximal subgroup of G. Then $N(M;G) = M$, and so $x^{-1}Mx \neq M$ for all $x \notin M$. Therefore, by Problem 1.43, $x^{-1}Mx \cap M$ is normal in $\langle x^{-1}Mx,M\rangle = G$, and so, by hypothesis, $x^{-1}Mx \cap M = 1$ for all $x \notin M$. If we write $|M| = m$ and $|G| = n$, then, by Problem 1.42, G contains exactly $(n/m) - 1$ elements which do not lie in any conjugate of M in G. Let u be one of these elements, and let K be a maximal subgroup of G containing u.

Since K does not lie in any conjugate of M, therefore, for all x and y in G, $\langle x^{-1}Kx,y^{-1}My\rangle = G$. Therefore, by Problem 1.43, $x^{-1}Kx \cap y^{-1}My$ is a proper normal subgroup of G, and so, by hypothesis, is equal to 1. If $|K| = k$, then, by Problem 1.42, K and its conjugates in G contain $n(k-1)/k$ nonidentity elements of G. None of these elements lie in any conjugate of M, so $n(k-1)/k \leqslant (n/m) - 1$; that is, $n + 1 \leqslant (n/m) + (n/k)$. Since m and k are both greater than 1, we have a contradiction.

1.45. We proceed by induction on the order of G. Let us suppose that G is a noncyclic group of order n. By the induction hypothesis, each proper subgroup is abelian. Therefore, by Problem 1.44, G has a normal subgroup N distinct from G or 1. Then N and G/N both satisfy the given conditions, and so, by induction, each is abelian. Let M be a maximal subgroup of G containing N. Since G/N is abelian, M is normal in G and G/M is cyclic. By induction, M is abelian, and so, by hypothesis, $M \subseteq Z(G)$. Thus G is abelian by Problem 1.15.

(For the results of Problems 1.42 to 1.45, see [26]. In that paper, the result of Problem 1.45 is used to prove the theorem: *Every finite division ring is a field.*)

1.46. Put $C = \langle xy\rangle$. Because $x^2 = y^2 = 1$, we have $x^{-1}(xy)x = y^{-1}x^{-1} = (xy)^{-1}$ and $y^{-1}(xy)y = y^{-1}x^{-1} = (xy)^{-1}$. Therefore $x^{-1}Cx = y^{-1}Cy = C$. Thus

C is normal in $\langle x,y\rangle = G$. Since $G = \langle C,x\rangle = C\langle x\rangle$, the index of C in G divides $|\langle x\rangle| = 2$. Since G is not cyclic, $C \neq G$, and so $|G:C| = 2$. If $|G| = 2k$, then it is readily verified that there is an isomorphism of G onto D_k taking $x \to a$ and $y \to ab$ (see Problem 1.6).

1.47. The set of all subgroups $\langle x^{-n}yx^n\rangle$ ($n = 0, \pm 1, \pm 2, \ldots$) has neither a maximal nor a minimal element since $\langle x^{-n}yx^n\rangle$ is a proper subgroup of $\langle x^{-n-1}yx^{n+1}\rangle$ for each n.

1.49. (b) Take H as the group of all matrices of the given form with $k = 0$ and such that the coefficients of $p(x)$ are all integers. Let

$$u = \begin{bmatrix} 2^{-1} & 0 \\ 0 & 1 \end{bmatrix}.$$

(This example is due to R. Steinberg.)

***1.50.** See [18], page 461. E. Landau has given a bound which implies that we may always take $\beta(r) \leqslant r^{2^r}$ (see [27]).

Solutions for Chapter 2—Permutation Groups

2.1. See [1], page 43.

2.2. Using 2.T.4, we find that S_5 has seven classes of conjugates which contain respectively: the identity 1, the ten 2-cycles, the twenty 3-cycles, the fifteen permutations of the form $(a\ b)(c\ d)$, the thirty 4-cycles, the twenty permutations of the form $(a\ b\ c)(d\ e)$, and the twenty-four 5-cycles. Similarly, A_5 has five classes, and the orders of these classes are respectively: 1, 20, 15, 12, and 12. A normal subgroup of a finite group G must be a union of complete classes of conjugates of G, and its order must divide the order of G. Thus we see by inspection that the normal subgroups of S_5 are 1, A_5, and S_5, and that A_5 is simple (see [9], page 49).

2.3. If $x \in N$, then, by 2.T.4, $x^{-1}(1\ 2\ \ldots\ n)x = (1^x\ 2^x\ \ldots\ n^x)$, where x maps the letter i onto i^x. Since $x^{-1}(1\ 2\ \ldots\ n)x$ generates $\langle(1\ 2\ \ldots\ n)\rangle$, it is equal to $(1\ 2\ \ldots\ n)^k$ for some integer k relatively prime to n. Thus $i^x + k \equiv$

$(i+1)^x \pmod{n}$ for $i = 1, 2, \ldots, n$. In particular, it follows that $i^x \equiv k(i-1) + 1^x \pmod{n}$. Thus x is a permutation of the form $i^x \equiv ki + l \pmod{n}$, where k and l are integers, $0 \leqslant k, l < n$, with k relatively prime to n. Conversely, any permutation of this form is in N. The order of N is $n\phi(n)$, where $\phi(n)$ is the Euler ϕ-function (compare [18], Section 140).

2.4. For further details, see [18], page 229; [1], page 79; and [5], page 285.

2.5. By induction on n. The result is trivial if $n = 2$, so let us suppose that $n > 2$. Let $x \in S_n$, and suppose that $n^x = m$. Then the permutation $x(1\ m)(1\ n)$ leaves n fixed and so lies in S_{n-1}. Thus, by the induction hypothesis, this latter permutation, and so x itself, is a product of the given transpositions.

2.6. By induction on n. The result is trivial for $n = 3$, so let us suppose that $n > 3$. Let $x \in A_n$, and suppose that $n^x = m$. Then $x(12m)(12n)^{-1}$ is an even permutation which leaves n fixed, and so lies in A_{n-1}. Hence, by induction, this latter permutation, and so x itself, is in the group G generated by the given 3-cycles. Then $A_n \subseteq G$. Since each 3-cycle is an even permutation, $G \subseteq A_n$, and so the result follows.

2.7. Since A_n is a normal subgroup of index 2 in S_n, and G is not contained in A_n, therefore $GA_n = S_n$. Hence $2 = |GA_n{:}A_n| = |G{:}A_n \cap G|$ by 1.T.5, and $A_n \cap G$ is the required normal subgroup of G.

2.8. We first note that

$$A = \bigcup_{n=1}^{\infty} A_n$$

is a normal subgroup of S by Problem 1.31. $A \neq S$ because $(12) \notin A$. On the other hand, for any $x \in S$, either x or $x(12)$ is an even permutation, and so x lies in either A or $A(12)$. Thus A has two cosets in S; that is, $|S{:}A| = 2$.

Now let us suppose that $N \neq 1$ is a normal subgroup of S. If $S \neq N$, then, for some n, $N \cap S_m \neq S_m$ or 1, for all $m \geqslant n$. We may take $n \geqslant 5$, and then, by 2.T.5, we see that $N \cap S_m = A_m$ for all $m \geqslant n$. Hence N contains precisely the even permutations of S; that is, $N = A$.

2.9. If $\gamma \in \alpha^G \cap \beta^G$, then $\gamma = \alpha^a$ for some $a \in G$. Therefore

$$\gamma^G = \{\alpha^{ax} \mid x \in G\} = \{\alpha^y \mid y \in G\} = \alpha^G.$$

Similarly, $\gamma^G = \beta^G$, and so $\alpha^G = \beta^G$.

2.10. If $x,y \in G_\alpha$, then $\alpha^x = \alpha^y = \alpha$. Therefore $\alpha^{x^{-1}} = \alpha^{xy} = \alpha$, and so x^{-1} and xy both lie in G_α. Then G_α is a subgroup of G, by 1.T.6.

2.11. We note that $u \in G_\beta$ if and only if $(\alpha^x)^u = \alpha^x$; that is, $\alpha^{xux^{-1}} = \alpha$, or equivalently, $xux^{-1} \in G_\alpha$. Hence $xG_\beta x^{-1} = G_\alpha$.

2.12. Let $x,y \in G$. Then $\alpha^x = \alpha^y$ if and only if $xy^{-1} \in G_\alpha$; that is, $G_\alpha x = G_\alpha y$. Hence the number of distinct elements in the orbit α^G is equal to the number of right cosets of G_α in G, which is $|G|/|G_\alpha|$ by 1.T.1 (see [7]).

2.13. Let $\Omega = \{Ax \mid x \in G\}$, and define $(Ax)^b = Axb$ for $b \in B$ and $Ax, Axb \in \Omega$. Then the double coset AxB is the union of the set of right cosets of A in the orbit $(Ax)^B$. Since, by Problem 2.9, Ω is a set union of disjoint orbits, G is a set union of disjoint double cosets of the form AxB. The stabilizer of Ax is $\{b \in B \mid Axb = Ax\} = \{b \in B \mid b \in x^{-1}Ax\} = B \cap x^{-1}Ax$. Hence, by Problem 2.12, the number of right cosets of A lying in the orbit $(Ax)^B$ is $|B|/|B \cap x^{-1}Ax|$, and so $|AxB| = |A|\,|B|/|B \cap x^{-1}Ax|$ as required.

2.14. The mapping is into S_Ω by (a), and a homomorphism by (b). The kernel of the homomorphism is $\{x \in G \mid \alpha^x = \alpha \text{ for all } \alpha \in \Omega\} = \bigcap_{\alpha \in \Omega} G_\alpha$.

2.15. Since $G_\alpha = 1$ for all $\alpha \in \Omega$, the result follows from Problem 2.14.

2.16. We proceed by induction on k. The result holds for $k = 0$, and so we assume that $k \geqslant 1$. Let z be an element of G having order 2^k; that is, z generates a subgroup of order 2^k of G. In the regular representation of G, z is mapped onto a permutation z^* with m cycles, each of length 2^k. A cycle of length 2^k is odd, by 2.T.2, and so, since m is odd, the permutation z^* is odd. Hence, by Problem 2.7, the image of the regular representation, which is isomorphic to G by Problem 2.15, has a normal subgroup of index 2. Hence G has a normal subgroup H of index 2. Since $z^2 \in H$, the conditions for the induction hypothesis are satisfied, and so we conclude that H has a normal subgroup N of index 2^{k-1} and order m.

We now assert that N is the only subgroup of G of order m. Let us suppose that M is a subgroup of order m in G. Since m is relatively prime to 2^k, it follows from Problem 1.17 that $M \subseteq H$, and so $M \subseteq N$; that is, $M = N$. Since N is the unique subgroup of order m in G, it is normal in G, and the proof of the result is complete.

2.17. The stabilizer of Ha is $\{y \in G \mid Hay = Ha\} = \{y \in G \mid y \in a^{-1}Ha\} = a^{-1}Ha$. Hence, by Problem 2.14, the kernel of the homomorphism is as described. If $N \subseteq H$, and N is a normal subgroup of G, then $N = a^{-1}Na \subseteq a^{-1}Ha$ for all $a \in G$, and so $N \subseteq K$ as required.

2.18. The homomorphism defined in Problem 2.17 gives a normal subgroup K of G in H such that G/K is isomorphic to a subgroup of S_n. Therefore $|G:K|$ divides $|S_n| = n!$.

2.19. Let K be defined as in Problem 2.18, with $n = p$. Then $|G:K|$ divides both $|G|$ and $p!$. Since p is the smallest prime dividing $|G|$, this implies that $|G:K|$ divides p. Hence $K = H$.

2.20. Let H be any proper subgroup of index $\leqslant n - 1$ in S_n, and let K be defined as in Problem 2.18, with $G = S_n$. Then $|G:K|$ divides $(n-1)!$, and so $K \neq 1$. Since K is a proper normal subgroup of S_n, this implies that $K = A_n$, by 2.T.5. Since $K \subseteq H \subset G$, and $|G:A_n| = 2$, therefore $H = A_n$.

2.21. We have

$$|\Omega| = \binom{p^r m}{p^r} = m\binom{p^r m - 1}{p^r - 1} = m\prod_{k=1}^{p^r-1}\left(\frac{p^r m}{k} - 1\right).$$

Since $p^r \nmid k$ for $1 \leqslant k < p^r$, we can write $p^r m/k = pm_k/l_k$, where m_k and l_k are relatively prime integers and $p \nmid l_k$. Multiplying out the product above, we have

$$\binom{p^r m - 1}{p^r - 1} = (-1)^{p^r-1} + \frac{pn}{l},$$

where p and n are integers which are relatively prime to the integer l. Since the binomial coefficient is an integer, $l = 1$. Therefore

$$|\Omega| = m\{(-1)^{p^r-1} + pn\} \equiv m \pmod{pm}.$$

2.22. Let Ω be as in Problem 2.21, with $S = G$. Define $\alpha^x = \{ax \mid a \in \alpha\}$ for $x \in G$ and $\alpha, \alpha^x \in \Omega$. Then G acts as a permutation group on Ω. Let $\alpha \in \Omega$, and suppose that $a \in G$ lies in α. Then the coset aG_α is contained in α. Hence α is a subset of p^r elements of G consisting of complete left cosets of G_α. In particular, $|G_\alpha|$ divides p^r. From Problem 2.12, the orbit α^G has length $|G:G_\alpha|$, which is equal to $p^s m$ for some integer $s \geqslant 0$. Furthermore, $s = 0$ if and only if $|G_\alpha| = p^r$; that is, α is a left coset aG_α, say, of G_α in G. Conversely, if α is a left coset aH of some subgroup H in G, then $H = G_\alpha$. There are mn_r left cosets of subgroups of order p^r in G, and these are all elements of Ω. For the remaining $\alpha \in \Omega$, $|G:G_\alpha| = |\alpha^G| \equiv 0 \pmod{pm}$, and so the number of elements of Ω which are not left cosets of subgroups of order p^r in G is congruent to 0 (mod pm). It follows that $|\Omega| \equiv mn_r$ (mod pm). Hence, by Problem 2.21, $m \equiv mn_r$ (mod pm); that is, $n_r \equiv 1$ (mod p). (See [28].)

2.23. Using Problem 2.13, we can write $|G| = |Px_1Q| + |Px_2Q| + \cdots + |Px_sQ|$ for some elements x_i in G. Let $p^r = |P|$. Then p^{r+1} does not divide $|G|$, and so p^{r+1} does not divide the order of some double coset $|PxQ|$, say. Since $|PxQ| = |P|\,|Q:x^{-1}Px \cap Q|$, by Problem 2.13, therefore p does not divide $|Q:x^{-1}Px \cap Q|$. Since Q is a p-group, this implies that Q is contained in $x^{-1}Px$, and hence $xQx^{-1} \subseteq P$.

2.24. A nonabelian group of order 21 has seven Sylow 3-groups and one Sylow 7-group, by using Problem 2.23. A nonabelian group of order 39 must have thirteen Sylow 3-groups and one Sylow 13-group.

2.25. If G were a simple group of order 56, then it would have eight Sylow 7-groups. Any two of these subgroups must have common intersection 1, and so between them the Sylow 7-groups contain $8 \cdot (7 - 1) = 48$ different elements of order 7. The remaining elements of G must then make up a unique (normal) Sylow 2-group in G. This shows that G is not simple.

2.26. Let us suppose that G were a simple noncyclic group of order $2^m p^n$, with $m = 1, 2$, or 3. Let P be a Sylow p-group of G; let Q be a Sylow 2-group of G; and put $N = N(P;G)$. Then, by Problem 2.23, either $|G:N| = 4$ and $p = 3$, or $|G:N| = 8$ and $p = 7$. In the first case, Problem 2.18, with $H = N$, implies that $|G|$ divides $4!$, and so $n = 1$. Thus, in this case, Q has index 3 in G, and another application of Problem 2.18, with $H = Q$, shows that G has order 6, and so is not simple. In the second case, Problem 2.18 shows that $|G|$ divides $8!$, and so $n = 1$. Thus $|G| = 56$, and Problem 2.25 shows that G is not simple in this case.

2.27. Let $\langle u \rangle$ be a Sylow 2-group of G, and let $\langle v \rangle$ be a Sylow 3-group. If both of these subgroups are normal in G, then G is their direct product, and so G is abelian. In this case, $G = \langle uv \rangle$, and so G is cyclic. Thus, if G is not cyclic, it follows from Problem 2.23 that $\langle v \rangle$ is normal in G, and so $\langle u \rangle$ is not. Then the homomorphism described in Problem 2.17, with $H = \langle u \rangle$, has kernel 1. Thus, because $\langle u \rangle$ has three cosets in G, G is isomorphic to a subgroup of S_3. Since $|G| = |S_3|$, therefore $G \simeq S_3$.

2.28. We first show that a Sylow 3-group $P = \langle x \rangle$ of G is not normal in G. Otherwise, x has at most two conjugates (x and x^2), and so $|G:C(x;G)| \leqslant 2$, by 1.T.3. Then some element y of order 2 commutes with x. If Q is a Sylow 2-group containing y, then Q is abelian (it has order 4), and so y lies in the center of $\langle Q,x \rangle = G$. This contradicts our hypothesis; so we conclude that P is not normal in G.

Finally, applying Problem 2.17 to G, with $H = P$, we find (as in Problem 2.27) that G is isomorphic to some subgroup of S_4. It is readily verified that A_4 is the only subgroup of order 12 in S_4, and so $G \simeq A_4$.

2.29. An equivalent assertion is: No permutation group which is generated by $n - 2$ transpositions has an orbit of length $\geqslant n$.

We shall prove this latter result by induction on n. Let us suppose that the contrary holds, and that G is a group with an orbit of length at least n and G is generated by $n - 2$ transpositions. Then one of the letters, say α, in the given orbit appears in exactly one of the given transpositions. We assume that $(\alpha\beta)$ is the transposition in which it occurs. The remaining $n - 3$ transpositions generate a group H whose elements leave α fixed. By induction, H has no orbit of length greater than $n - 2$. Since $(\alpha\beta)$ has at most one letter (namely, β), in common with any orbit of H, the group $G = \langle H,(\alpha\beta) \rangle$ cannot have any orbit of length greater than $(n - 2) + 1 = n - 1$. This is the required contradiction (see [29]).

2.30. The stabilizers of G are conjugate by Problem 2.11; and Ω is the only orbit for G, so $| G:G_\alpha | = | \Omega | = n$, by Problem 2.12.

2.31. If G is not primitive, then let Γ be a nontrivial block. Let H be the subgroup of all elements in G which map Γ into itself. It is readily verified that H is a proper subgroup of G, and that H properly contains G_α for any $\alpha \in \Gamma$. Thus G_α is not a maximal subgroup of G, and hence neither are any of the other (conjugate) stabilizers of G.

Conversely, if H is a proper subgroup of G, and H properly contains some stabilizer G_α of G, then H is intransitive on Ω. Let Γ be the orbit α^H of H. Then we assert that Γ is a nontrivial block for G. First, $\alpha^H \neq \{\alpha\}$ because $H \neq G_\alpha$. Second, if $x \in G$ and $\Gamma^x \cap \Gamma \neq \varnothing$, then, for some β and γ in Γ, we have that $\beta^x = \gamma$. But, for some $u \in H$, $\beta^u = \gamma$, and so $ux^{-1} \in G_\beta \subseteq H$. Therefore $x \in H$, and so $\Gamma^x \cap \Gamma = \Gamma$. Thus Γ is a nontrivial block for G, and G is not a primitive group.

2.32. If G is an intransitive group (different from 1), then G has an orbit of length at least 2. This orbit is a nontrivial block for G.

2.33. Let us suppose that a normal subgroup N of G lies in some stabilizer G_α of G. Then, since the stabilizers of G are all conjugate, N lies in every stabilizer of G. Thus $N = 1$. Therefore, if G contains a nontrivial normal subgroup N, N is not contained in G_α. Since G_α is a maximal subgroup of G, by Problem 2.31, therefore $G = G_\alpha N$. Hence $\Omega = \alpha^G = \alpha^{G_\alpha N} = \alpha^N$, and so N is transitive.

2.34. Let H be the subgroup consisting of all elements of G which map Γ into itself. Let $\alpha \in \Gamma$; then $G_\alpha \subseteq H$. Since G is transitive, we can find elements u_β $(\beta \in \Gamma)$, which must lie in H, such that u_β takes α into β. If $x \in H$, then x takes α into β if and only if $u_\beta x^{-1}$ lies in G_α; that is, $G_\alpha u_\beta = G_\alpha x$. Therefore G_α has exactly $|\Gamma|$ right cosets in H. Since $n = |G:G_\alpha| = |G:H|\,|H:G_\alpha|$, by 1.T.1, therefore $|\Gamma| = |H:G_\alpha|$ divides n.

2.35. Let $x|\Omega_\lambda$ denote the restriction of the element x in G to Ω_λ. It is readily verified that the mapping $x \to x|\Omega_\lambda$ $(x \in G)$ is a homomorphism of G onto $G|\Omega_\lambda$. The kernel of the homomorphism is $\{x \in G \mid \alpha^x = \alpha \text{ for all } \alpha \in \Omega_\lambda\} = N$.

2.36. From Problem 2.35, $G|\Omega_i \simeq G/N_i$ for $i = 1, 2, \ldots, s$. Since

$$\bigcap_{i=1}^{s} N_i = \bigcap_{\alpha \in \Omega} G_\alpha = 1,$$

the result follows from Problem 1.19.

2.37. By Problem 2.12, the lengths of the orbits of G divide p^k. Since no orbit has length greater than $n < p^2$, all orbits of G have length 1 or p. Let us suppose that Ω_λ is an orbit of length p for G. If $\alpha \in \Omega_\lambda$, then $|G:G_\alpha| = p$, and so, by Problem 2.19, G_α is normal in G. Thus, by Problem 2.30, $G_\alpha = 1$.

In general, we can state that, on each orbit Ω_λ, $G|\Omega_\lambda$ is either the identity group (when $|\Omega_\lambda| = 1$) or else a group of order p (when $|\Omega_\lambda| = p$). Therefore the direct product in Problem 2.36 is an elementary abelian p-group, and the result follows.

2.38. Since G is regular, any stabilizer $G_\alpha = 1$. Since G is transitive, $|G| = |G:G_\alpha| = n$, by Problem 2.30.

2.39. Using the notation of Problem 2.15, we note that the image of the regular representation is transitive because, for any $a,b \in \Omega$, the element $a^{-1}b$ in G maps a into b. Furthermore, for $a \in \Omega$, $G_a = \{x \in G \mid ax = a\} = 1$, and so the permutation group is regular.

2.40. Let G be a transitive abelian group on a set Ω. For each $\alpha \in \Omega$, the stabilizer G_α is normal in G. Since the stabilizers of G are all conjugate in G, by Problem 2.30, therefore $G_\alpha = 1$ for all $\alpha \in \Omega$.

2.41. By induction on n. An abelian group G of degree n is either transitive or intransitive. In the former case, $|G| = n \leqslant 3^{n/3}$, by Problem **2.40.** In the latter case, we suppose that G has orbits $\Omega_1, \Omega_2, \ldots, \Omega_s$. Then, by the induction hypothesis, $|G|\Omega_i| \leqslant 3^{n_i/3}$, where $n_i = |\Omega_i|$. Therefore, by Problem 2.36, $|G| \leqslant 3^{n_1/3}3^{n_2/3}\cdots 3^{n_s/3} = 3^{n/3}$. [*Note:* There exist abelian groups of order 3^m and degree $3m$ for $m = 1, 2, \ldots$]

2.42. If $x \in Z(G)$ fixes a letter α, say, then the normal subgroup $X = \langle x \rangle$ of G lies in G_α. Since G is transitive, all the stabilizers of G are conjugate, by Problem 2.30, and so the normal subgroup X lies in each of the stabilizers of G. Thus x fixes all letters; that is, $x = 1$.

Since $|G:G_\alpha| \geqslant |G_\alpha Z(G):G_\alpha| = |Z(G):Z(G) \cap G_\alpha| = |Z(G)|$, it follows from Problem 2.30 that $n \geqslant |Z(G)|$.

2.43. (a) Choose α_i as an element of Ω_i for $i = 1, 2, \ldots, k$. Since G is transitive, there exist elements x_i in G such that $\alpha_1^{x_i} = \alpha_i$ for $i = 1, 2, \ldots, k$. For all $u \in N$, $\Omega_1^{x_i u} = \Omega_1^{x_i}$, and so $\Omega_1^{x_i}$ is mapped into itself by elements of N. Since $\alpha_i \in \Omega_1^{x_i}$, the orbit Ω_i containing α_i lies in $\Omega_1^{x_i}$. Similarly, since $\alpha_i^{x_i^{-1}} = \alpha_1$, we find that Ω_1 lies in $\Omega_i^{x_i^{-1}}$; that is, $\Omega_1^{x_i}$ lies in Ω_i. The result then follows. (b) From (a) we conclude that all orbits have the same order. Since the total of the lengths of the orbits is n, each orbit has length n/k. (c) The mapping $u|\Omega_1 \to (x_i^{-1}ux_i)|\Omega_i$ $(u \in N)$, where x_i is defined as in (a), gives the required isomorphism. (This is analogous to a theorem of Clifford. See Problem 10.8.)

2.44. From Problem 2.31, it is necessary and sufficient that the group G should have 1 as a maximal subgroup, that is, that 1 should be the only proper subgroup of G. This is true if and only if G has prime order.

2.45. By Problem 2.44, either G has prime order or G is not regular. In the latter case, G_α and G_β are nontrivial maximal subgroups of G, by Problem 2.31, and so $G = \langle G_\alpha, G_\beta \rangle$.

2.46. By Problem 2.45, we may consider two cases. If G is regular, then the result is trivial; so let us suppose that $G = \langle G_\alpha, G_\beta \rangle$. In this case, using Problems 1.43 and 2.30, we have that $N = G_\alpha \cap G_\beta$ is a normal subgroup of G lying in G_α. Hence $N = 1$, by Problem 2.30.

2.47. The result is trivial if G is regular; so we may suppose that the stabilizers of G are nontrivial. Then, by Problem 2.45 and 1.T.2, $|G| = |G:1| \leqslant |G:G_\alpha||G:G_\beta| \leqslant p^2$. Since G is a subgroup of S_p, its order divides $p!$, and so, using Problem 2.30, we conclude that p divides $|G|$

and that $|G| < p^2$. Therefore, from Problem 2.23, the Sylow p-group (of order p) in G has its normalizer equal to G; that is, it is a normal subgroup of G.

2.48. By Problem 2.47, the Sylow p-group P of G is normal and of order p. Clearly, P is generated by a p-cycle u, say, and so $G_\gamma \cap P = 1$ for all $\gamma \in \Omega$. We have to show that, if $x,y \in G_\gamma$, then $x^{-1}y^{-1}xy = 1$. In fact, since P is normal, there exist integers r and s such that $xux^{-1} = u^r$ and $yuy^{-1} = u^s$. Then $x^{-1}y^{-1}xyu = ux^{-1}y^{-1}xy$, and so the subgroup $A = \langle P, x^{-1}y^{-1}xy \rangle$ is abelian. Since A is a transitive abelian permutation group of degree p, $|A| = p$, by Problem 2.40. Hence $x^{-1}y^{-1}xy \in P \cap G_\gamma = 1$, and the result follows (see [7], Theorem 11.6).

2.49. A transitive p-group has as its degree a power of p, by Problem 2.30. Therefore each orbit of P has length p^k for some integer k. The result now follows from Problem 2.43(b).

2.50. Let G_α be a stabilizer for G. Using Problem 2.33, we find that $G = G_\alpha Z(G)$, and so G_α is normal in G. Therefore $G_\alpha = 1$, by Problem 2.30, and hence G has prime order, by Problem 2.44.

2.51. Let G_α be a stabilizer of G. Because of Problem 2.31, it is sufficient to show that G_α is a maximal subgroup of G. Let us suppose that, on the contrary, G_α is not a maximal subgroup of G. Then there is a proper subgroup K of G properly containing G_α. Since $K = K \cap G_\alpha N = G_\alpha$, some $x \neq 1$ in N lies in K. Therefore $K \cap N = M$, say, is a nontrivial subgroup of G. Moreover, M is normal in K because N is normal in G, and M is normal in N because N is abelian. Therefore M is normal in $KN = G$. Since N is a minimal normal subgroup of G, this implies that $M = N$, and so $N \subseteq K$. Therefore $K = G$, which is the required contradiction.

2.52. The number of k-cycles in S_n is $\binom{n}{k}(k-1)!$. If $|G| \geqslant (n-k)!k$, then $|S_n{:}G| \leqslant n!/[(n-k)!k] = \binom{n}{k}(k-1)!$. Hence either G contains a k-cycle x or some right coset of G contains at least two k-cycles y and z. In the latter case we take $x = yz^{-1}$. (This result is due to Netto.)

2.53. Let us consider the orbits of P operating as a permutation group on the set Ω in the manner defined. Since a transitive p-group is of degree p^k for some integer $k \geqslant 0$, all the orbits of P must have lengths which are powers of p: $l_1 = 1, l_2, \ldots, l_k$, say. ($l_1$ is the length of the orbit $\{1\}$.)

Since p divides $|P| = l_1 + l_2 + \cdots + l_k$, it follows that at least p orbits must have length 1. The elements of P which lie in these orbits must be in $Z(P)$.

2.54. Let P be a Sylow p-group and Q be a Sylow q-group of G. From Problems 2.22 and 2.23, $|G:N(P:G)| \equiv 1 \pmod{p}$. But $N(P;G) = P$ or G. Since $|G:P| = q \not\equiv 1 \pmod{p}$, by hypothesis, therefore $N(P;G) = G$; that is, P is normal in G. Thus P is the unique Sylow p-group of G, and G has $p - 1$ elements of order p. If $x \in P$ has order p, then $C(x;G)$ is equal to either P or G. Since every conjugate of x has order p, the number of conjugates of x (which is $|G:C(x;G)|$ by 1.T.3) is less than q. Therefore $C(x;G) = G$. Hence $P \subseteq Z(G)$, and G is abelian by Problem 1.15.

2.55. Since n is relatively prime to $\phi(n) = n \prod_{p|n} (1 - 1/p)$, therefore $n = p_1 p_2 \cdots p_k$, where the p_i are distinct prime numbers such that $p_i \nmid p_j - 1$ for all i,j. We shall proceed by induction on k. Since the result is true for $k = 1, 2$, from Problem 2.54, we shall suppose that $k \geqslant 3$.

By the induction hypothesis, each proper subgroup of G is abelian. Therefore, by Problem 1.44, G has a normal subgroup $N \neq G$ or 1. By induction, G/N is abelian, so all its subgroups are normal. Thus a maximal subgroup M/N of G/N has prime index, say, p_k, in G/N. Then the group M is a normal abelian subgroup of index p_k in G, from the induction hypothesis. If $i \neq k$, then M has a unique Sylow p_i-group $P_i \neq 1$, from Problem 2.23. For each $x \in G$, we have $x^{-1}P_ix \subseteq x^{-1}Mx = M$, and so $x^{-1}P_ix$ is a Sylow p_i-group of M; that is, $x^{-1}P_ix = P_i$. Hence P_i is normal in G for $i = 1, 2, \ldots, k - 1$. By the induction hypothesis, G/P_1 and G/P_2 are abelian groups, and so the external direct product $G/P_1 \times G/P_2$ is also abelian. The result now follows from Problem 1.19 because $P_1 \cap P_2 = 1$. (This result is due to Burnside. It follows from 3.T.6 that G is cyclic, and so all groups of the given order are isomorphic.)

2.56. By Problem 2.53, $G/Z(G)$ has order 1 or p and is therefore cyclic. Thus G is abelian by Problem 1.15.

2.57. Now V has order 168. We may take $P = \langle(1\ 2\ 3\ 4\ 5\ 6\ 7)\rangle$ and verify that $P^V = V$. By Problem 2.34, V is primitive, and so Problem 2.33 shows that any nontrivial normal subgroup N of V has order divisible by 7. Thus, by Problem 1.17, $P \subseteq N$, and so $N = N^V \supseteq P^V = V$. This shows that V is simple.

2.58. A group of order pm, with p prime and $m < p$, has a normal Sylow p-group, by Problem 2.23. This observation, together with Problems 2.53

and 2.26, allows us to conclude that any simple group of order $\leqslant 100$ must have one of the orders: $2^5 \cdot 3$, $2^4 \cdot 3$, $2^4 \cdot 5$, $2^2 \cdot 3 \cdot 5$, $2 \cdot 3 \cdot 5$, $3^2 \cdot 5$, $3^2 \cdot 7$, or $3 \cdot 5^2$. An obvious application of Problem 2.18, with H as an appropriate Sylow subgroup, reduces the list to: $2^2 \cdot 3 \cdot 5$, $2 \cdot 3 \cdot 5$, or $3^2 \cdot 7$. The cases $2 \cdot 3 \cdot 5$ and $3^2 \cdot 7$ are eliminated by Problems 2.16 and 2.23, respectively, and so the result follows (see also [18], page 504).

***2.59.** See [18], Section 127.

***2.60.** There are only seven such groups (up to a natural isomorphism), and their degrees are 4, 6, 6, 8, 12, 12, and 24, respectively (see [18], Chapter 12).

***2.61.** See [18], Section 166.

***2.62.** This result is due to C. Jordan. See [7], Theorem 13.9.

***2.63.** See [30].

***2.64.** See [2], Section 54. A number of interesting examples appear in [31].

Solutions for Chapter 3—Automorphisms and Finitely Generated Abelian Groups

3.1. There are 2, 3, 5, and 6 groups, respectively.

3.2. There are two nonisomorphic abelian groups. Any nonabelian group of order 12 is isomorphic to exactly one of the following: the alternating group A_4, the dihedral group D_6 (see Problem 1.6), and a group generated by elements x and y subject only to the relations

$$y^6 = 1, \qquad x^{-1}yx = y^{-1}, \qquad \text{and} \qquad x^2 = y^3.$$

(Compare [18], Section 59.)

3.3. There are two groups in each case. If $p = 2$, then each nonabelian group of order 8 is isomorphic to one of the groups Q, of Problem 1.4, or T, of Problem 1.5. For $p \neq 2$, each nonabelian group of order p^3 is isomorphic to one of the following: (a) a group generated by elements x and y

subject only to the relations $x^p = y^{p^2} = 1$, $x^{-1}yx = y^{p+1}$; (b) a group generated by elements x, y, and z subject only to the relations $x^p = y^p = z^p = 1$, $x^{-1}yx = yz = zy$, $xz = zx$ (see [18], Sections 117 to 118 or [9], page 151).

3.4. We may proceed by induction on the number m of cyclic subgroups occurring in the decomposition of 3.T.5, and suppose that $m > 1$. Let $h_1 > 1$ be the smallest order of any subgroup occurring in such a decomposition. Then $G = C_1 \times H$, where C_1 is a cyclic subgroup of order h_1, and H is a direct product of $m - 1$ cyclic subgroups. By the induction hypothesis, $H = C_2 \times \cdots \times C_m$, where the order h_i of C_i divides h_{i+1} for $i = 2, \ldots, m - 1$. We have to show $h_1 \mid h_2$.

Let d be the greatest common divisor of h_1 and h_2, and put $n_i = h_i/d$ $(i = 1, 2)$. Since n_1 and n_2 have greatest common divisor 1, there exist integers s and t such that $n_1s - n_2t = 1$. Put $C_1^* = \langle z_1^{n_1}z_2^{n_2}\rangle$ and $C_2^* = \langle z_1^{t}z_2^{s}\rangle$, where z_1 and z_2 generate C_1 and C_2, respectively. Then $C_1^* \times C_2^* = C_1 \times C_2$, and C_1^* has order d. Since $G = C_1^* \times C_2^* \times \cdots \times C_m$, the choice of h_1 implies that $d = h_1$; that is, $h_1 \mid h_2$.

3.5. If $n = 1$, the result follows from 3.T.4. Therefore we may proceed by induction on n. Let us suppose that $x_1, \ldots, x_n$ generate G, and that H is a subgroup of G. Then we choose $y = x_1^{m_1} \cdots x_n^{m_n}$ as an element of H such that the exponent m_1 has the smallest possible nonzero value. (If x_1 never appears to a nonzero power in any element of H, then $\langle x_2, \ldots, x_n\rangle$ contains H and the result follows from the induction hypothesis.) Taking y^{-1} if necessary, we may assume that $m_1 > 0$. Now for any $z = x_1^{l_1} \cdots x_n^{l_n}$ in H, we can find integers q and r, with $0 \leqslant r < m_1$, such that $l_1 = qm_1 + r$. Then the power of x_1 in zy^{-q} is r, and by the choice of m_1 this means $r = 0$. Hence $H = \langle y,K\rangle$, where $K = H \cap \langle x_2, \ldots, x_n\rangle$. By the induction hypothesis, K is generated by a set of $n - 1$ or fewer elements; so the result follows.

3.6. Any two rational numbers have integral multiples which are equal. Therefore any two nontrivial subgroups of A will have a nontrivial intersection. In particular, the result as stated is true.

3.7. Trivial.

3.8. Let x generate the cyclic group C of order n. If α is an automorphism of C, then, by Problem 3.7 and 3.T.2, $x^\alpha = x^k$ for some integer k relatively prime to n with $1 \leqslant k \leqslant n$. Conversely, if k is relatively prime to n, then the mapping $u \to u^k$ $(u \in C)$ is an automorphism of C. Let R_n be the multiplicative group of residue classes of the integers modulo n relatively prime to n. If $\bar{k}$ denotes the residue class (mod n) which contains the integer k, then we can define a mapping θ of Aut C into R_n by

$$\theta : \alpha \to \bar{k} \qquad \text{if and only if } x^\alpha = x^k.$$

It is readily verified that θ is an isomorphism of Aut C onto R_n. Since R_n is an abelian group of order $\phi(n)$, the same is true for Aut C.

Now let $C_1 = \langle y \rangle$ be a cyclic group of infinite order. Then the only generators of C_1 are y and y^{-1}, by 3.T.2. Thus, by Problem 3.7, every automorphism of C_1 maps y into either y or y^{-1}. Thus C_1 has exactly two automorphisms, namely, $u \to u$ (for all $u \in C_1$) and $u \to u^{-1}$ (for all $u \in C_1$).

3.9. Each automorphism of S_3 permutes the set $\{(12), (23), (13)\}$, and no nontrivial automorphism leaves each element of the set fixed. Therefore, if two automorphisms α and β permute the set in the same way, then $\alpha\beta^{-1} = 1$ and $\alpha = \beta$. Hence S_3 has at most $3! = 6$ automorphisms. Since $Z(S_3) = 1$, S_3 has 6 inner automorphisms by 3.T.1, and so Aut $S_3 \simeq S_3$.

3.10. Consider the four Sylow 3-groups $P_1 = \langle(234)\rangle$, $P_2 = \langle(134)\rangle$, $P_3 = \langle(124)\rangle$, and $P_4 = \langle(123)\rangle$ of S_4. By Problem 2.23, each automorphism of S_4 permutes the set Ω of these subgroups. Furthermore, by 2.T.4, $x^{-1}P_ix = P_j$ when the permutation $x \in S_4$ maps i onto j. Thus, an element $\alpha \in$ Aut S_4 which maps each P_i onto itself must also leave each $x \in S_4$ fixed, and is therefore trivial. Hence different automorphisms of S_4 correspond to different permutations of Ω, and so $|\text{Aut } S_4| \leqslant |S_\Omega| = 24$. Since $Z(S_4) = 1$, S_4 has 24 inner automorphisms. Thus Aut $S_4 \simeq S_4$, and each automorphism is an inner automorphism, by 3.T.1. (The results of Problems 3.9 and 3.10 are special cases of a general theorem. See [18], Section 162.)

3.11. Cyclic groups of orders 3 and 4 and the infinite cyclic group all have automorphism groups of order 2. The noncyclic group of order 4 and the symmetric group S_3 have automorphism groups isomorphic to S_3.

3.12. First, let us suppose that G is a group with two generators x and y. We may choose x and y so that the order m of x divides the order n of y, using Problem 3.4. Then define three mappings of G into itself by:

$$\alpha\colon x^s y^t \to x^{s+t} y^t, \qquad \beta\colon x^s y^t \to x^s y^{-t}, \qquad \gamma\colon x^s y^t \to x^t y^s.$$

Because $m \mid n$, α is always a well-defined, nontrivial automorphism of G. If $n \neq 2$, then β is also a nontrivial automorphism of G, and $\alpha\beta \neq \beta\alpha$. If $n = 2$, then β is the identity mapping, but in this case γ is a well-defined automorphism of G, and $\alpha\gamma \neq \gamma\alpha$. Thus Aut G is nonabelian in this case.

In the general case, we may suppose that G is a product of cyclic subgroups as in Problem 3.4. Then Aut G contains a subgroup consisting of all automorphisms of G which leave the elements of $C_3, \ldots, C_m$ fixed. This subgroup is isomorphic to Aut $(C_1 \times C_2)$, which we have already shown to be nonabelian. (This result is due to G. A. Miller.)

3.13. If α is an automorphism of G, then, for all $x \in G$, x and x^α have the same order. Hence, by the hypothesis on G, all the nontrivial elements of G have the same order. If a prime p divides $|G|$, then G has an element of order p, by Problem 2.22, and so all nontrivial elements of G have order p. Furthermore, if $G \neq 1$, then $Z(G) \neq 1$, by Problem 2.53. But then, for any $x \neq 1$ in $Z(G)$, we have $x^\alpha \in Z(G)$ for each $\alpha \in$ Aut G, because $Z(G)$ is a characteristic subgroup of G. Thus, by the hypothesis on G, $Z(G) = G$; that is, G is abelian (see [9], page 55).

3.14. If α denotes the given automorphism, then, for all $x,y \in G$, $(xy)^\alpha = (xy)^{-1} = y^{-1}x^{-1} = y^\alpha x^\alpha = (yx)^\alpha$. Hence $xy = yx$.

3.15. If $x,y \in G$ are such that $x^{-1}x^\alpha = y^{-1}y^\alpha$, then $(xy^{-1})^\alpha = x^\alpha(y^\alpha)^{-1} = xy^{-1}$. Hence $xy^{-1} = 1$; that is, $x = y$. Thus there are $|G|$ distinct elements of the form $x^{-1}x^\alpha$ $(x \in G)$, and so they must constitute all of G.

3.16. By Problem 3.15 we may write each $z \in G$ in the form $x^{-1}x^\alpha$ for some $x \in G$. Then $z^\alpha = (x^{-1})^\alpha x^{\alpha^2} = (x^\alpha)^{-1}x = z^{-1}$, and so G is abelian, by Problem 3.14. Finally, $z^{-1} = z^\alpha \neq z$ for all $z \neq 1$ in G. Therefore G has no element of order 2, and so $|G|$ is odd, by Problem 2.22. (See [18], Section 66. A deep generalization was proved by Thompson [32]. Also see [4], Section 9.4.)

3.17. For each $a \in N(H;G)$, the mapping $\bar{a}: x \to a^{-1}xa$ $(x \in H)$ is an automorphism of H, namely, the restriction to H of the corresponding inner automorphism of $N(H;G)$. It is readily verified that the mapping $a \to \bar{a}$ $(a \in N(H;G))$ is a homomorphism of $N(H;G)$ onto some subgroup A, say, of Aut H. Since $\bar{a} = 1$ if and only if $ax = xa$ for all $x \in H$, the kernel of the homomorphism is $C(H;G)$. Therefore $N(H;G)/C(H;G) \simeq A$.

3.18. From Problems 3.17 and 3.8.

3.19. For each $x \in G$, $x^{-1}Px \subseteq x^{-1}Hx = H$, and so $x^{-1}Px$ is a Sylow p-group of G. By Problem 2.23, $x^{-1}Px = u^{-1}Pu$ for some $u \in H$, and so $xu^{-1} \in N(P;G)$. Thus $x \in N(P;G)H$ for all $x \in G$; that is, $G = N(P;G)H$.

3.20. The inner automorphisms of G, when restricted to the normal subgroup H, are automorphisms of H. Since K is a characteristic subgroup of H, it is left fixed by all inner automorphisms of G; that is, K is normal in G. In the case when H is characteristic in G, every automorphism of G, when restricted to H, is an automorphism of H. Hence, in this case, K is left fixed by every automorphism of G; that is, K is characteristic in G.

3.21. We first show that, if $y,z \in G$ and $y^n = z^n$, then $y = z$. Let $m = |G|$. Since m and n are relatively prime, there exist integers s and t such that $ms + nt = 1$. Then $y = y^{ms+nt} = y^{nt} = z^{nt} = z^{ms+nt} = z$ because the orders of y and z both divide m.

It now follows that the set $\{y^n \mid y \in G\}$ contains $|G|$ distinct elements of G, and so comprises the whole of G. Thus $x = y^n$ for some unique y in G.

3.22. Since m is relatively prime to n, $G = AH = AK$, by Problem 1.8, and $A \cap H = A \cap K = 1$. Thus H and K are each complete sets of coset representatives for A in G. Therefore we can define a one-to-one mapping θ of H onto K by the condition $Ax = Ax^\theta$ $(x \in H; x^\theta \in K)$. Since $A(xy)^\theta = Axy = (Ax)(Ay) = (Ax^\theta)(Ay^\theta) = A(x^\theta y^\theta)$ for any $x,y \in H$, it follows that θ is an isomorphism of H onto K.

We now construct c. Since $u^\theta u^{-1} \in A$ for all $u \in H$, and A is abelian, we can define $b \in A$ by $b = \prod_{u \in H} u^\theta u^{-1}$. For all $x \in H$, we have

$$\begin{aligned} xbx^{-1} &= \prod_{u \in H} \{x(x^\theta)^{-1}(xu)^\theta(xu)^{-1}\} = \{x(x^\theta)^{-1}\}^n \prod_{u \in H} \{(xu)^\theta(xu)^{-1}\} \\ &= \{x(x^\theta)^{-1}\}^n b. \end{aligned}$$

Because m is relatively prime to n, we can use Problem 3.21 to find $c \in A$ such that $b = c^n$. Then $(xcx^{-1})^n = xc^nx^{-1} = \{x(x^\theta)^{-1}c\}^n$. Thus, by Problem 3.21, $xc^{-1}x^{-1} = x(x^\theta)^{-1}c$; that is, $x^\theta = cxc^{-1}$. In particular, $K = cHc^{-1}$ (see [9], page 163).

3.23. We first note that H is the only Hall subgroup of order $|H|$ in $N(H;G)$. For, if K is a subgroup of that order in $N(H;G)$, then $K = K \cap N(H;G) = K \cap H$, by Problem 1.13, and so $K = H$. Hence H is a characteristic subgroup of $N(H;G)$. If N is the normalizer of $N(H;G)$ in G, then $N(H;G)$ is normal in N, and so, by Problem 3.20, H is normal in N. Thus $N \subseteq N(H;G)$, and the result follows.

3.24. Let us suppose that, for some $x \in G$, $x^{-1}Tx = S$. Let $M = N(S;G)$. Then both P and $x^{-1}Px$ lie in M. Hence, by Problem 2.23, there exists $u \in M$ such that $u^{-1}Pu = x^{-1}Px$, and so $xu^{-1} = y$, where $y \in N(P;G)$. Then we have $y^{-1}Ty = ux^{-1}Txu^{-1} = uSu^{-1} = S$ as required.

3.25. By hypothesis, each element of P lies in a distinct conjugacy class of $N(P;G)$, so the result follows from Problem 3.24 (see [18], Section 123).

3.26. We have $x = u^{-1}yu$ if and only if $x^\alpha = v^{-1}y^\alpha v$ where $u^\alpha = v$, for any x,y,u, and v in G.

3.27. Let $\alpha \in A$ and $\beta \in \text{Aut } G$. We have to show that each set C of conjugates in G is mapped into itself by $\beta^{-1}\alpha\beta$. This follows from $C^{\beta^{-1}\alpha\beta} = \{(C^{\beta^{-1}})^\alpha\}^\beta = (C^{\beta^{-1}})^\beta = C$, using Problem 3.26.

3.28. Because of the Sylow theorem in Problem 2.22, it is sufficient to show that, if $\alpha \in A$ has prime order p, then p divides $|G|$. Let H be the subset of those elements in G which are each mapped onto themselves by α. Now H is clearly a subgroup of G, and so, by Problem 1.9, not every element in G is conjugate to an element of H. Hence there exists a conjugacy class C in G which is disjoint from H. Since α maps C into itself, α restricted to C is a permutation of the set of elements in C. Since α has order p, the lengths of the orbits under $\langle\alpha\rangle$ in C divide p, by Problem 2.12. Since $C \cap H = \varnothing$, this implies that C is a union of a number of disjoint orbits each of length p. In particular, p divides $|C|$, and so p divides $|G|$, by 1.T.3 (see [33], Section 40 and [18], Section 65).

3.29. Let D be a subgroup of smallest order with the property that, for some $x \in G$, $D = x^{-1}Px \cap P$. We have to show that $D = 1$. Let us suppose that $D \neq 1$, and let $H = \langle x^{-1}Px,P\rangle$. By Problem 1.43, D is normal in H, and so $H \neq G$, by hypothesis. Since D is not normal in G, there exists a Sylow p-group P_1 of G such that P_1 does not contain D. It then follows from Problem 2.23 that P_1 does not lie in H, but that $P_1 \cap H \subseteq u^{-1}Pu$ for some $u \in H$. Then $u^{-1}x^{-1}Pxu \cap P_1 = u^{-1}x^{-1}Pxu \cap H \cap P_1 \subseteq u^{-1}x^{-1}Pxu \cap u^{-1}Pu = u^{-1}Du = D$. Since P_1 does not contain D, this implies that $u^{-1}x^{-1}Pxu \cap P_1$ is a proper subgroup of D. This contradicts the choice of D.

3.30. From Problem 3.29 we may find a Sylow p-group P of G, and an element x in G, such that $P \cap x^{-1}Px = 1$. Let $N = N(P;G)$. Then $N \cap x^{-1}Px = 1$, by Problem 1.13, and so the elements $x^{-1}Px$ must lie in distinct cosets of N. In particular, $|G:N| \geqslant p^n$ and, since p does not divide $|G:N|$, therefore $|G:N| \geqslant p^n + 1$. Hence P has at least $p^n + 1$ conjugates in G. Finally, $|G| = |G:N|\,|N| \geqslant (p^n + 1)p^n$. (For the results of Problems 3.29 and 3.30, see [34].)

3.31. Using Problem 3.29, we may choose x in G such that $P \cap x^{-1}Px = 1$. Then, by Problem 2.13, the double coset PxP contains $|P|$ right cosets of P. In the given permutation representation for P, the orbit Ω_1 containing the element Px is the set of right cosets for P in PxP. The group $P|\Omega_1$ is a transitive permutation group of degree $|P|$, and so has order $\geqslant |P|$, by Problem 2.30. Since this permutation group is a homomorphic image of P and contains as many elements as P, the representation is faithful.

3.32. Let $N_1, N_2, \ldots, N_m$ be the distinct normal subgroups of G with indices $\leqslant n$. Using Problem 2.17, we can find a homomorphism θ_i of G into the symmetric group S_n with kernel N_i for $i = 1, 2, \ldots, m$. If $x_1, x_2, \ldots, x_k$ is a set of generators for G, then the homomorphism θ_i is completely defined by the values in S_n of the images of these generators under θ_i. In particular, the number of homomorphisms of G into S_n is not greater than the number of possible sets of values for these k image elements, and so is at most $(n!)^k$. Since N_i is the kernel of θ_i, and $N_i \neq N_j$ for $i \neq j$, it follows that $m \leqslant (n!)^k$ as asserted.

3.33. It follows from Problem 1.7 that H contains at least one normal subgroup N of G which is of finite index in G. By Problem 3.32 there are only a finite number of normal subgroups in G of the same index. Let us suppose that they are $N = N_1, N_2, \ldots, N_h$. Any automorphism of G clearly permutes this set of normal subgroups, and so

$$K = \bigcap_{i=1}^{h} N_i$$

is a characteristic subgroup of G. Finally, using 1.T.2, it follows that K is of finite index in G (see [35]).

3.34. It follows from 3.T.5 that G is a direct product of n cyclic groups of order p. If $\mathcal{V}$ is a vector space of dimension n over the field $\mathcal{F}$ of integers (mod p), then $\mathcal{V}$ is an additive abelian group which is clearly isomorphic to G. Hence Aut $G \simeq$ Aut $\mathcal{V}$.

A mapping α of $\mathcal{V}$ into itself is an automorphism if and only if α has an inverse mapping and (a) $(u + v)^\alpha = u^\alpha + v^\alpha$; (b) $(mu)^\alpha = mu^\alpha$; for each $u,v \in \mathcal{V}$ and each integer m. The conditions (a) and (b) are exactly the conditions that α be a linear transformation of $\mathcal{V}$. Thus Aut $\mathcal{V}$ is the group of all invertible linear transformations of $\mathcal{V}$ onto itself. From elementary matrix theory it is known that this latter group is isomorphic to the group of all nonsingular $n \times n$ matrices with entries in $\mathcal{F}$. This proves the result (see [18], Sections 89 to 90).

3.35. Let us consider the number of ways of choosing a nonsingular $n \times n$ matrix with entries in the field of integers (mod p). In such a matrix the first row can be any nonzero n-vector and so may be chosen in $p^n - 1$ ways. In general, for $1 < k \leqslant n$, if the first $k - 1$ rows have been already chosen, then the kth row must be chosen to be linearly independent of the preceding rows. This means that it must not lie in the $(k-1)$th-dimensional subspace spanned by the first $k - 1$ rows, and so it may be chosen in $p^n - p^{k-1}$ ways. Hence there are $(p^n - 1)(p^n - p) \cdots (p^n - p^{n-1})$ ways of choosing different $n \times n$ matrices with entries in the given field. The stated result now follows from Problem 3.34.

3.36. Because of Problem 3.34, it is sufficient to show that in the group of all nonsingular $n \times n$ matrices with entries in the field of integers (mod p) all nontrivial p-elements have order p.

Let $x = [\xi_{ij}]$ be an $n \times n$ matrix over the field of integers (mod p) such that $x^{p^k} = 1$ for some $k \geqslant 1$. Then the minimal equation for x divides $\xi^{p^k} - 1 = (\xi - 1)^{p^k}$, and so all the eigenvalues of x equal 1. We may therefore transform x into Jordan form which will consist of blocks of the form

$$y = \begin{bmatrix} 1 & 1 & 0 & \cdots & 0 \\ 0 & 1 & 1 & \cdots & 0 \\ 0 & 0 & 1 & \cdots & 0 \\ \cdot & & & & \cdot \\ \cdot & & & & \cdot \\ \cdot & & & & \cdot \\ 0 & 0 & 0 & \cdots & 1 \end{bmatrix}.$$

By induction on $s \geqslant 1$, we find that

$$y^s = \begin{bmatrix} 1 & \eta(s,1) & \eta(s,2) & \cdots & \eta(s,m-1) \\ 0 & 1 & \eta(s,1) & \cdots & \eta(s,m-2) \\ 0 & 0 & 1 & \cdots & \eta(s,m-3) \\ \cdot & & & & \cdot \\ \cdot & & & & \cdot \\ \cdot & & & & \cdot \\ 0 & 0 & 0 & \cdots & 1 \end{bmatrix}$$

where $\eta(s,i)$ is the congruence class (mod p) which contains $s!/[i!(s-i)!]$ if $i \leqslant s$, and is 0 otherwise.

For $s = p$, we have $p!/[i!(p-1)!] \equiv 0 \pmod p$ for $1 \leqslant i \leqslant p - 1$. Since $m \leqslant n \leqslant p$, this implies that $\eta(p,i) = 0$ for all i, and so $y^p = 1$. Since this is true for each of the blocks in the Jordan form for x, $x^p = 1$.

(More generally, if G is an elementary abelian p-group of order p^n with $p^{r-1} < n \leqslant p^r$, then the greatest order of any p-element of Aut G is p^r. For some related results see [36] and [37].)

3.37. (b) In this case, G has order $p(p-1)$, and the matrices whose diagonal entries are equal to 1 form a characteristic subgroup of order p. Put

$$u = \begin{bmatrix} 1 & 1 \\ 0 & 1 \end{bmatrix}, \quad v = \begin{bmatrix} 1 & 0 \\ 0 & a \end{bmatrix}, \quad \text{and} \quad w = \begin{bmatrix} 1 & 0 \\ 0 & 2 \end{bmatrix},$$

and suppose that $\alpha \in \text{Aut } G$. Then we can write

$$u^\alpha = \begin{bmatrix} 1 & k \\ 0 & 1 \end{bmatrix} \quad \text{and} \quad v^\alpha = \begin{bmatrix} 1 & t(a) \\ 0 & a' \end{bmatrix}$$

for some integer k, $1 \leqslant k \leqslant p-1$, and some integers $t(a)$ and a' with $p \nmid a'$. Applying α to the identities $uv = vu^a$ and $vw = wv$, we find that $a' = a$ and $t(a) = (a-1)t(2)$. Thus, by direct calculation, $x^{-1}ux = u^\alpha$ and $x^{-1}vx = v^\alpha$, where

$$x = \begin{bmatrix} 1 & -t(2) \\ 0 & k \end{bmatrix}.$$

3.38. Let G be the group defined in Problem 1.47. Then the subgroup H consisting of all matrices with diagonal entries equal to 1 is not finitely generated. (By [6] every group of countable order may be embedded as a subgroup in a group generated by two elements.)

3.39. Let G be the permutation group $\langle(12), (135)(246)\rangle$. Then $H = \langle(12), (34), (56)\rangle$ is a normal subgroup of order 8 and index 3 in G, and G has seven elements of order 2 and eight elements of order 3. Thus, G has at most $7 \cdot 8 = 56$ sets of generators consisting of an element of order 2 and one of order 3. Hence, using Problem 3.7, we conclude that $|\text{Aut } G| \leqslant 56$. However, H is an elementary abelian group of order 8, and so $|\text{Aut } H| = 7 \cdot 6 \cdot 4 = 168$, by Problem 3.35.

***3.40.** See [9], page 148. Compare with Problems 3.10 and 3.11.

***3.41.** See [18], Sections 310, 311, and 327. A complete discussion of these and similar groups is found in [38]. See also [3], Chapter 8, and [4], pages 116 to 123. Further related results appear in [39], [40], and [41]. A recent survey of the known finite simple groups is given in [42].

Solutions for Chapter 4—Normal Series

4.2. Using the notation of Problem 1.6, we see that the subgroup $\langle b \rangle$ of D_k is the only cyclic subgroup of order k, and so it is a characteristic subgroup. Thus, if $\alpha \in \text{Aut } D_k$, then $a^\alpha = ab^i$ and $b^\alpha = b^j$ for some integers i and j, with j relatively prime to k because $\langle b^\alpha \rangle = \langle b \rangle$. Conversely, for any integers i and j, with j relatively prime to k, the mapping of D_k into itself defined by $a^s b^t \to (ab^i)^s (b^j)^t$ $(s = 0, 1; t = 0, 1, \ldots, k-1)$ is an automorphism (by direct calculation). Thus $\text{Aut } D_k = H_0 K_0$, where K_0 is the set of all automorphisms which leave a fixed, and H_0 is the set of all automorphisms which leave b fixed. We have $H_0 \cap K_0 = 1$; H_0 is a normal cyclic subgroup of order k; and K_0 is an abelian group of order $\phi(k)$, by Problem 3.8. It is now straightforward to show that $\text{Aut } D_k$ is isomorphic to a semidirect product of H_0 by K_0.

4.3. Let G be the group S defined in Problem 2.8, and let H be the infinite abelian subgroup generated by the transpositions

$$(2m - 1\ 2m) \qquad (m = 1, 2, \ldots).$$

Then G has the composition series $S \supset A \supset 1$; but any normal series

$$H = H_0 \supset H_1 \supset \cdots \supset H_n = 1$$

for H has a proper refinement. (See also the result quoted in Problem *1.50.)

4.4. Take G as the permutation group generated by (12), (56), and (13)(24). Then $H = \langle (12), (34), (56) \rangle$ is a normal subgroup of G, and we have the composition series

$$G \supset H \supset \langle (12), (56) \rangle \supset \langle (12) \rangle \supset 1,$$

which has the property required.

4.5. (b) Clearly G has the principal series: $G \supset T_0 \supset 1$, with $T_0 \simeq T$ and $G/T_0 \simeq A$. Since each normal series for T_0 has a proper refinement, G has no composition series.

4.6. Let G be the infinite cyclic group $\langle x \rangle$. Then two such descending chains in G are

$$\langle x \rangle \supset \langle x^2 \rangle \supset \langle x^4 \rangle \supset \langle x^8 \rangle \supset \cdots, \quad \text{and} \quad \langle x \rangle \supset \langle x^3 \rangle \supset \langle x^9 \rangle \supset \langle x^{27} \rangle \supset \cdots.$$

(Generalizations of normal series are considered in [2], Section 56.)

4.7. Let $|G| = p^n$. We shall proceed by induction on n. By Problem 2.53, $Z(G) \neq 1$, for $n \geqslant 1$, and so, by Problem 2.22, there exists a subgroup N of order p in $Z(G)$ which is clearly normal in G. By induction there is a principal series

$$G/N \supset G_1/N \supset \cdots \supset G_{n-1}/N = 1$$

for G/N with all its normal factors of order p. Then

$$G \supset G_1 \supset \cdots \supset G_{n-1} = N \supset 1$$

is a principal series for G.

4.8. By induction on $|G|$. By Problem 2.53, $Z = Z(G) \neq 1$. If $Z \subseteq H$, then H/Z is a proper subgroup of G/Z, and induction shows that $N(H;G)Z/Z = N(H/Z;G/Z) \supset H/Z$; that is, $N(H;G) \supset H$. On the other hand, if H does not contain Z, then $H \subset HZ \subseteq N(H;G)$.

4.9. Analogous to Problem 1.37.

4.10. Let H be a proper normal subgroup of G. Then we may construct a proper ascending chain, $H = H_0, H_1, \ldots,$ of normal subgroups of G as follows. Given H_i ($i = 0, \ldots, k - 1$), either H_{k-1} is a maximal normal subgroup of G (and we are finished) or we can choose H_k as a proper normal subgroup of G properly containing H_{k-1}. By hypothesis the chain is finite, and the last term will be a maximal normal subgroup of G containing H. The second part is analogous.

4.11. If G satisfies the normal chain condition, then we can construct a composition series, $H_0 = G, H_1, \ldots,$ for G as follows. Given H_i ($i = 0, \ldots, k - 1$), either $H_{k-1} = 1$ (and we are finished) or we can choose a maximal proper normal subgroup H_k of H_{k-1}, by Problems 4.9 and 4.10. By hypothesis the chain is finite; so $H_n = 1$ for some n, and we have a composition series for G.

Conversely, if G has a composition series of length l, then any finite descending chain of type (a) in the definition of normal chain condition yields a normal series

$$G = G_0 \supset G_1 \supset \cdots \supset G_k \supseteq 1$$

for G. This series may be refined to a composition series of length l by 4.T.2, and so $k \leqslant l$. Thus each chain of this type has length at most l. Similarly, each ascending chain of type (b) has length at most l. Thus G satisfies the normal chain condition.

4.12. If K is not simple, then, by Problems 4.9 and 4.10, we may find a minimal normal subgroup N of K. For each $x \in G$, $x^{-1}Nx$ is also a minimal normal subgroup of G. We define $M_0 = N$. Choosing $x_1 \in G$ such that $x_1^{-1}Nx_1$ does not lie in M_0, we define $M_1 = \langle M_0, x_1^{-1}Nx_1\rangle$. More generally, given M_k ($k \geqslant 0$) different from K, we choose x_{k+1} in G such that $x_{k+1}^{-1}Nx_{k+1}$ does not lie in M_k. We then define $M_{k+1} = \langle M_k, x_{k+1}^{-1}Nx_{k+1}\rangle$. Since $M_0, M_1, \ldots,$ is a proper ascending chain of normal subgroups of K, it must terminate at, say, M_n. Then it follows that M_n is normal in G because it is generated by the set of all elements in G conjugate to elements in N. Hence $M_n = K$, because K is a minimal normal subgroup of G. Furthermore, $M_{k-1} \cap x_k^{-1}Nx_k = 1$ for each k, because it is a normal subgroup of K properly contained in $x_k^{-1}Nx_k$. Therefore we have

$$K = M_n = N \times x_1^{-1}Nx_1 \times \cdots \times x_n^{-1}Nx_n.$$

Finally, each normal subgroup of N is centralized by each of the other direct factors of K, and hence such a subgroup must be normal in K. Since N is a minimal normal subgroup of K, this implies that N is simple.

4.13. Using Problem 4.10, we may choose a minimal normal subgroup N_1 of G. Then we choose subgroups $N_2, N_3, \ldots,$ of G successively such that

N_i/N_{i-1} is a minimal normal subgroup of G/N_{i-1} ($i = 2, 3, \ldots$). Because G satisfies the normal chain condition, the chain $1 \subset N_1 \subset N_2 \subset \cdots$ is finite and hence must terminate with $N_k = G$ for some k. This yields a principal series for G.

Finally, let

$$G = G_0 \supset G_1 \supset \cdots \supset G_n = 1$$

be any principal series for G. By definition, G_i/G_{i+1} is a minimal normal subgroup of G/G_{i+1} for $i = 1, 2, \ldots, n-1$. The result then follows from Problem 4.12.

4.14. Let

$$G \supset A_1 \supset A_2 \supset \cdots \supset A_m = A \qquad \text{and} \qquad G \supset B_1 \supset B_2 \supset \cdots \supset B_n = B$$

be normal series from G to A and G to B, respectively. Then

$$G \supset A_1 \supset \cdots \supset A_m = A \supseteq A \cap B_1 \supseteq A \cap B_2 \supseteq \cdots \supseteq A \cap B_n = A \cap B$$

is a normal series from G to $A \cap B$. (For this, and some of the following results on subnormal subgroups, see the fundamental paper [43].)

4.15. Let

$$G = G_0 \supseteq G_1 \supseteq \cdots \supseteq G_n = H$$

be any normal series from G to H. We shall show by induction that $G_i \supseteq H_i$ ($i = 1, 2, \ldots, m$), and so $n \geqslant m$ as asserted. In fact, $G_0 = H_0$, and if $G_{i-1} \supseteq H_{i-1}$, then $H_i = H^{H_{i-1}} \subseteq H^{G_{i-1}} \subseteq G_i$.

4.16. If H has the minimal normal series given in Problem 4.15, then the series

$$G = H_0^\alpha \supset H_1^\alpha \supset \cdots \supset H_m^\alpha = H^\alpha$$

is a normal series for H^α. In fact, we have $H_i^\alpha = (H^\alpha)^{H_{i-1}^\alpha}$, and so, from Problem 4.15, $m(G,H) = m(G,H^\alpha)$.

4.17. We proceed by induction. If $H_{i-1}^\alpha = H_{i-1}$, then

$$H_i^\alpha = (H^\alpha)^{H_{i-1}^\alpha} = H^{H_{i-1}} = H_i.$$

4.18. We define $B_0 = G$ and $B_i = B^{B_{i-1}}$ ($i = 1, 2, \ldots$). If $m = m(G,B)$, then $B_m = B$, from Problem 4.15. Since the automorphism $x \to a^{-1}xa$ ($x \in G$) of G leaves B fixed for each $a \in A$, it follows from Problem 4.17 that $a^{-1}B_i a = B_i$ for $i = 1, 2, \ldots, m$. In particular, $\langle A,B_i \rangle = AB_i$ for each i, by 1.T.5.

We now proceed by induction on i to show that AB_i is subnormal in G. Let

$$G \supset A_1 \supset A_2 \supset \cdots \supset A_n = A$$

be a normal series from G to A. Then

$$AB_{i-1} \supseteq A_1 \cap AB_{i-1} \supseteq \cdots \supseteq A_n \cap AB_{i-1} = A$$

is a normal series from AB_{i-1} to A. Since B_i is normal in AB_{i-1}, therefore

$$AB_{i-1} \supseteq B_i(A_1 \cap AB_{i-1}) \supseteq \cdots \supseteq B_i(A_n \cap AB_{i-1}) = AB_i$$

is a normal series from AB_{i-1} to AB_i. Thus, if AB_{i-1} is subnormal in G, then AB_i is subnormal in G, and the induction step is proved. The result now follows.

4.19. There is a minimal normal series

$$H \supset M_1 \supset \cdots \supset M_n = M$$

from H to M. Let us suppose that M is not normal in H. Then $n \geqslant 2$, and, for some $x \in M_{n-2}$, we have $x^{-1}Mx \neq M$. Since $x^{-1}Mx \subseteq M_{n-1} \subseteq N(M;G)$, it follows that $Mx^{-1}Mx$ is a subnormal subgroup of G, by Problem 4.18. Since $Mx^{-1}Mx$ is a subgroup of H properly containing M, we have the required contradiction.

4.20. Put $H = \langle A,B \rangle$. Then, by the hypothesis on G, we may find a subgroup M which is maximal in the set of subnormal subgroups of G contained in H. By Problem 4.19, M is normal in H. Therefore, by Problem 4.18, AM and BM are subnormal subgroups of G. By the choice of M, AM and BM are both contained in M, and so $H = \langle A,B \rangle = M$. Thus H is subnormal in G.

Finally, we suppose that G has a composition series of length n, say. Using 4.T.2, we deduce that, for each subnormal subgroup S of G, the length $l(G,S)$ of the longest proper normal chain from G to S is not greater than n. Then, if $\mathbb{S}$ is any nonempty set of subnormal subgroups of G, any element S of $\mathbb{S}$ for which $l(G,S)$ is least is obviously a maximal element. (Problems 4.19 and 4.20 are unpublished results of H. Wielandt.)

4.21. Let

$$G = H_0 \supset H_1 \supset \cdots \supset H_n = H$$

be a minimal normal series from G to H. If $n > 1$, then there exists an element $x \in H_{n-2}$ such that $x^{-1}Hx \neq H$. But $x^{-1}Hx \subseteq H_{n-1} \subseteq N(H;G)$, and so it follows from Problem 1.13 that $x^{-1}Hx = H \cap x^{-1}Hx$. This contradicts the choice of x.

4.22. Let $m = m(G,A)$. The result holds if $m = 1$, since in this case A is normal and $\langle A,B \rangle = AB$. We proceed by induction on m. Let $B = \{b_1, b_2, \ldots, b_n\}$. Since $b_i^{-1}Ab_i$ lies in A^G for each i, we have, by Problem 4.16, that $m(A^G, b_i^{-1}Ab_i) = m - 1$. Therefore, by the induction hypothesis, we find successively that the subgroups $A_1 = b_1^{-1}Ab_1$ and $A_i = \langle A_{i-1}, b_i^{-1}Ab_i \rangle$ $(i = 2, 3, \ldots, n)$ are all finite subgroups of A^G. In particular, A_n is finite, and so $\langle A,B \rangle = A_nB$ is finite.

4.23. Put $H = \langle A,B \rangle$. By Problem 4.22, H is finite, and we proceed by induction on $| H:A |$ to show that A is normal in H. Since A is subnormal in H, there is a minimal normal series

$$H = A_0 \supset A_1 \supset \cdots \supset A_n = A$$

from H to A. Let us suppose that A is not normal in H. Then there exists $x \in A_{n-2}$ such that $x^{-1}Ax \neq A$. Since $x^{-1}Ax$ and A are both normal subgroups of A_{n-1}, $C = Ax^{-1}Ax$ is a normal subgroup of A_{n-1}. Furthermore, $| C | = | AxA | = | A | \, | A:x^{-1}Ax \cap A |$ (see Problem 2.13), and so C is a subnormal subgroup of H which properly contains A and whose order is relatively prime to $| B |$. Hence, by the induction hypothesis, C is normal in H. But then $H = BC$, and so B is a subnormal Hall subgroup of H. By Problem 4.21, B is normal in H, and so $H = AB$. Again A is a subnormal Hall subgroup of H, and so A is normal in H. Since the orders of A and B are relatively prime, $A \cap B = 1$. Thus H is the direct product of A and B (see [43]).

4.24. Let $H = \langle A,B \rangle$. By Problem 4.22, $\mathcal{S}$ as in Problem 4.19 is finite and so contains a maximal element M. The remainder of the proof is now analogous to the proof of Problem 4.20.

Solutions for Chapter 5—Commutators and Derived Series

5.1. Compute.

5.2. Let N be the smallest normal subgroup of G containing $[K,K]$. Since $[K,K] \subseteq G'$, therefore $N \subseteq G'$. On the other hand, if $x,y \in G$, then $x = u_1u_2 \cdots u_m$ and $y = v_1v_2 \cdots v_n$ for some $u_i, v_j \in K \cup K^{-1}$. We shall use induction on $m + n$ to show that $[x,y] \in N$. For $m + n \leqslant 2$, the result is easy. Let us suppose that $m + n \geqslant 3$, so that m, say, is greater than 1. Then, by Problem 5.1(a), $[x,y] = u_m^{-1}[u_1u_2 \cdots u_{m-1}, y]u_m[u_m, y]$, which lies in N by the induction hypothesis. Thus $[x,y] \in N$ for all x and y in G, and the result follows.

5.3. (a) Let $N = [[K,L],H][[L,H],K]$. Then N is normal in G, and so it follows from Problem 5.1(c) that $[[x,y],z] \in N$ for all $x \in H$, $y \in K$, and $z \in L$. By use of Problem 5.1(a), this implies that $[[H,K],z] \subseteq N$, and so the required result follows.

(b) For all $x \in H$, $y \in K$, and $z \in L$, it follows from Problem 5.1(a) that $[xy,z] \in y^{-1}[H,L]y[K,L] = [H,L][K,L]$ because H and L are normal subgroups. Hence $[HK,L] \subseteq [H,L][K,L]$. Since the reverse inequality is obvious, the result follows.

5.4. By Problem 5.1(a), $z^{-1}[x,y]z = [xz,y][z,y]^{-1} \in [H,K]$ for all $x,z \in H$ and $y \in K$. Hence $[H,K]$ is normalized by H. Similarly, $[H,K]$ is normalized by K, and so $[H,K]$ is a normal subgroup of $\langle H,K \rangle$.

5.5. Let $u,v \in H$ and $x \in G$. Then $[x^{-1}ux,v] = [uw,v] = w^{-1}[u,v]w[w,v] = [u,v]$, where $w = [u,x] \in G' \subseteq C(H;G)$. Similarly, $[u,x^{-1}vx] = [u,v]$. Because $G' \subseteq C(x^{-1}Hx;G)$, it follows that, for all $x \in G$, $x^{-1}[u,v]x = [x^{-1}ux,x^{-1}vx] = [x^{-1}ux,v] = [u,v]$; that is, $[u,v] \in Z(G)$. The result now follows.

5.6. Because x commutes with $[x,y] = x^{-1}(y^{-1}xy)$, therefore $[x,y]^m = x^{-m}(y^{-1}xy)^m = [x^m,y] = 1$. Similarly, $[x,y]^n = 1$. For some integers h and k, we have $mh + nk = d$. Hence $[x,y]^d = [x,y]^{mh}[x,y]^{nk} = 1$.

5.7. Let us suppose that the set $K = \{u,v\}$ generates G. Then, by hypothesis, $[K,K] = \langle [u,v] \rangle$ centralizes both u and v. Hence $[K,K] \subseteq Z(G)$, and the result follows from Problem 5.2.

5.8. By hypothesis, $_n[xux^{-1},u] = 1$ for all $x \in G$. Therefore

$$1 = x^{-1}(_n[xux^{-1},u])x = {_n}[u,x^{-1}ux] = {_{n-1}}[u,[u,x^{-1}ux]] = {_{n+1}}[u,x],$$

because $[u,x^{-1}ux] = u^{-1}(x^{-1}u^{-1}xu)u(u^{-1}x^{-1}ux) = {_2}[u,x]$ (see [44]).

5.9. Take G as the symmetric group S_3, and take $u = (123)$ and $y = (12)$. (This example is due to H. Heineken.)

5.10. We suppose that $G' \simeq S_4$. For each $x \in G$, the mapping $u \to x^{-1}ux$ ($u \in G'$) is an automorphism of G', and so, by Problem 3.10, there exists $v \in G'$ such that $x^{-1}ux = v^{-1}uv$ for all $u \in G'$. Therefore $xv^{-1} \in C(G';G) = C$, say, and so $x \in Cv \subseteq CG'$ for all $x \in G$. Thus $G = CG'$. However, $C \cap G' = Z(G') = 1$ because $G' \simeq S_4$. Since C is a normal subgroup of G, by Problem 1.12, this implies that G equals the direct product $C \times G'$. Since $C \simeq G/G'$, $C' = 1$, and so $G' = C' \times G'' = G''$. This gives a contradiction because the derived group of S_4 is $A_4 \neq S_4$ (see [18], Section 70).

5.12. For all $u \in N$ and all $x \in G$, $u^{-1}(x^{-1}ux) \in N \cap G' = 1$. Therefore $ux = xu$. Hence $N \subseteq Z(G)$.

5.13. Since $[G,A]$ is a normal subgroup of G, by Problem 5.4, therefore $[G,A]A$ is a subgroup. The normal closure A^G is the subgroup of G generated by the set $K = \{x^{-1}ax \mid x \in G \text{ and } a \in A\}$. On the other hand, $[G,A]A$ is generated by the union of the sets $H = \{x^{-1}a^{-1}xa \mid x \in G$ and $a \in A\}$ and A. Since $K \subseteq HA$, and A and H both lie in KK, the result follows.

5.14. Consider the chain, $G_0, G_1, \ldots,$ of subgroups of G defined by

$$G_0 = G \qquad \text{and} \qquad G_i = A^{G_{i-1}} \qquad (i = 1, 2, \ldots).$$

We assert that $G_i = A_iA$ for each i. This is true for $i = 0$, and, given $G_{i-1} = A_{i-1}A$, then $G_i = A^{G_{i-1}} = A^{A_{i-1}A} = [A,A_{i-1}]A = A_iA$, using Problem 5.13. Therefore, by induction, $G_i = A_iA$ for all i. The result now follows from Problem 4.15.

5.15. For each subgroup H of G we define the sequence of subgroups $H_0, H_1, \ldots,$ as in Problem 5.14. It then follows by induction on i that $(A \cap B)_i \subseteq A_i \cap B_i$ $(i = 0, 1, \ldots)$. For, if we suppose that $(A \cap B)_{i-1} \subseteq A_{i-1} \cap B_{i-1}$, then

$$\begin{aligned}(A \cap B)_i = [A \cap B,(A \cap B)_{i-1}] &\subseteq [A \cap B, A_{i-1} \cap B_{i-1}] \\ &\subseteq [A,A_{i-1}] \cap [B,B_{i-1}] = A_i \cap B_i.\end{aligned}$$

The result now follows from Problem 5.14.

5.16. Put $a_i = {}_{n-i}[x,a]$ for $i = 0, 1, \ldots, n-1$. All a_i lie in A and therefore commute with one another. We shall show by induction that $a_i^{h^i} = 1$. This is true for $i = 0$, by hypothesis, so let us suppose that $0 < i \leqslant n-1$, and that the identity holds for $i-1$. Then, because a_i commutes with $a_{i-1} = [x,a_i]$, $1 = [x,a_i]^{h^{i-1}} = [x,a_i^{h^{i-1}}]$. (Compare with the proof of Problem 5.6.) Thus x commutes with $a_i^{h^{i-1}} = [x,a_{i+1}]^{h^{i-1}} = [x,a_{i+1}^{h^{i-1}}]$. Therefore $a_i^{h^i} = [x,a_{i+1}^{h^{i-1}}]^h = [x^h,a_i^{h^{i-1}}] = 1$. This completes the induction proof, and for $i = n-1$ we have the stated result (see [45]).

5.17. Let $y \in H$. Then ${}_k[x,y] \in H$ for all nonnegative integers k, because H is normal. Also, ${}_n[x,y] = 1$ for some integer $n \geqslant 1$, because x is a nil-element. We assert that ${}_{n-i}[x,y] \in \langle x \rangle$ for $i = 0, 1, \ldots, n$. The result is trivial for $i = 0$, and, if ${}_{n-i+1}[x,y] \in \langle x \rangle$, then $(x^{-1}{}_{n-i}[x,y]x)^{-1}{}_{n-i}[x,y] \in \langle x \rangle$. Therefore ${}_{n-i}[x,y] \in N(\langle x \rangle;H) \subseteq \langle x \rangle$, so the induction step is proved. In particular, ${}_0[x,y] = y$ lies in $\langle x \rangle$.

5.18. Let P be a Sylow p-group containing x. Since G has more than one Sylow p-group, we can choose $u \in G$ such that $u^{-1}Pu \neq P$, by Problem

2.23. Let us consider the sequence $u_i = {}_i[x,u]$ $(i = 0, 1, \ldots)$. Since x is a nilelement, $u_k = 1$ for some integer $k \geqslant 1$. Thus, for some integer n between 0 and k, we have $u_n = v$, say, such that $v \notin N(P;G)$ and $u_{n+1} = [x,v] \in N(P;G)$. Then

$$vPv^{-1} = vxPx^{-1}v^{-1} = xv[x,v]^{-1}P[x,v]v^{-1}x^{-1} = xvPv^{-1}x^{-1}.$$

Therefore $x \in N(vPv^{-1};P)$ which equals $vPv^{-1} \cap P$ by Problem 1.13. The result follows.

5.19. If P is a Sylow p-group of G containing D, then $D \neq N \cap P$ by Problem 4.8. By Problem 2.23, N has a Sylow p-group Q such that $N \cap P \subseteq Q$. Similarly, if P_1 is a different Sylow p-group of G containing D, there is a Sylow p-group Q_1 of N such that $D \subset N \cap P_1 \subseteq Q_1$. Since Q and Q_1 are each contained in Sylow p-groups of G, by Problem 2.23, and $D \subseteq Q \cap Q_1$, therefore $Q \cap Q_1 = D$ by the choice of D. In particular, $Q_1 \neq Q$.

5.20. We use induction on $|G|$ to show that G has only one Sylow p-group. Let us suppose that G has more than one Sylow p-group. Then we define D as in Problem 5.19 and note that $D \neq 1$ by Problem 5.18. Since $N(D;G)$ has more than one Sylow p-group, by Problem 5.19, the induction hypothesis implies that $G = N(D;G)$; that is, D is normal in G. The p-elements of G/D are clearly all nilelements, and so by induction G/D has only one Sylow p-group. But this implies that G has only one Sylow p-group, contrary to our supposition. This proves the result. (See [46]. Related results appear in [4], Section 6.8; [47]; [48]; and the survey [49].)

5.21. It is sufficient to show that two commutators $[x,y]$ and $[u,v]$ in G are equal whenever x and u, and y and v, respectively, lie in the same cosets of $Z(G)$. But, if x and u lie in the same coset of $Z(G)$, then $xu^{-1} \in Z(G)$. Similarly, if y and v lie in the same coset of $Z(G)$, then $yv^{-1} \in Z(G)$. Then

$$x^{-1}y^{-1}xy = u^{-1}(xu^{-1})^{-1}v^{-1}(yv^{-1})^{-1}(xu^{-1})u(yv^{-1})v = u^{-1}v^{-1}uv$$

as required.

5.22. Since $G/Z(G)$ has order n, $[x,y]^n \in Z(G)$. Therefore

$$\begin{aligned}[x,y]^{n+1} &= x^{-1}y^{-1}x[x,y]^n y = x^{-1}y^{-2}xy^2y^{-1}[x,y]^{n-1}y \\ &= [x,y^2][y^{-1}xy,y]^{n-1}.\end{aligned}$$

5.23. From Problem 5.21, we have that there are at most n^2 distinct commutators in G. It is therefore sufficient to show that if $u \in G'$, and if u can be written as a product $c_1c_2 \cdots c_m$ of m commutators in G, and if one commutator $[x,y]$, say, occurs more than n times in the product, then u can be written as a product of $m - 1$ commutators. But, as in the proof of Problem 1.12, we can collect the terms in $[x,y]$ so that $u = [x,y]^{n+1}c'_{n+2} \cdots c'_m$, where each c'_i is a commutator conjugate to one of the commutators c_j under some power of $[x,y]$. Then, by Problem 5.22, $[x,y]^{n+1}$ is a product of n commutators in G; so u is a product of $m - 1$ commutators.

5.24. Since there are at most n^2 distinct commutators in G, by Problem 5.21, it follows from Problem 5.23 that there are at most $(n^2)^{n^s} = n^{2n^s}$ elements in G' (see [50]).

5.25. Each conjugate $u^{-1}xu$ lies in xG'. Hence x has at most $|\, xG' \,| = m$ conjugates in G.

5.26. Let us suppose that $x_1, x_2, \ldots, x_k$ is a set of generators for G. Then, by Problem 5.25 and 1.T.3, $|\, G{:}C(x_i\,;G) \,| \leqslant m$ for each i. But

$$Z = \bigcap_{i=1}^{k} C(x_i;G),$$

and so $|\, G{:}Z \,| \leqslant \prod_{i=1}^{k} |\, G{:}C(x_i\,;G) \,| \leqslant m^k$, using 1.T.2. (For stronger results, see [51] and [52].)

5.27. We proceed by induction on s. The result is true for $s = 1$. If $(xy)^s = x^sy^sz^{s(s-1)/2}$, then $(xy)^{s+1} = (xy)^sxy = x^{s+1}[x,y^{-s}]y^{s+1}z^{s(s-1)/2}$ because z commutes with x and y. But $[x,y^{-s}] = (x^{-1}yx)^sy^{-s} = z^s$ because $x^{-1}yx = zy$ and zy commutes with y. Thus $(xy)^{s+1} = x^{s+1}y^{s+1}z^{s(s+1)/2}$, and the induction step is proved.

5.28. We proceed by induction on $|\, G \,|$. The result is true when $|\, G \,| = p$, so we suppose that $|\, G \,| \geqslant p^2$. Now G possesses at least one subgroup of index p, by Problem 2.22, and all subgroups of index p are normal in G, by Problem 2.19. We suppose that G is not cyclic. Then it possesses at least two subgroups of index p (see Problem 1.11). By the induction hypothesis, these two subgroups are cyclic, say $\langle x \rangle$ and $\langle y \rangle$. By 1.T.2, the subgroup $N = \langle x \rangle \cap \langle y \rangle$ has index p^2 in G. Moreover, it is clear from 5.T.1 that $G' \subseteq N = Z(G)$ because $G = \langle x,y \rangle$. Thus, for any $u,v \in G$, $[u,v]^p = (u^{-1}v^{-1}u)^pv^p = u^{-1}v^{-p}uv^p = 1$, because $[u,v]v^{-1}$ commutes with v, and $v^p \in Z(G)$. Therefore it follows from Problem 5.27 that $(uv)^p = u^pv^p$ for all $u,v \in G$ because p is odd. This implies that the mapping $u \to u^p$ $(u \in G)$ is a homomorphism of G into $Z(G)$. Since $|\, G \,| = p^2 \,|\, Z(G) \,|$, the kernel K of this homomorphism has order $\geqslant p^2$. But all the nontrivial elements of K have order p, and so this contradicts the hypothesis that G has only one subgroup of order p. (See [18], Sections 104 to 105, or [9], pages 148 to 149, for a more general statement. A recent generalization is given in [53].)

***5.30.** This result is due to R. Dedekind. See [54], page 114 or [1], Theorem 12.5.4.

***5.31.** The following are the only nonsolvable groups of order $\leqslant 200$. There is a simple group of order 60, namely, the alternating group A_5. There are three nonisomorphic nonsolvable groups of order 120, and each of these has a composition factor isomorphic to A_5. There is a simple group of order 168 (see Problem 2.57), and one nonsolvable group of order 180 (see [55], Section 74).

***5.32.** This is a result of V. Dlab generalizing an earlier theorem of F. Szasz (see [56]). The proof hinges on the following result: If S is a subset of G, and $S(n)$ and $S(m)$ are both contained in $N(S;G)$, then if $S(m)$ and $S(n)$ are both abelian groups, $S(d)$ is abelian.

Solutions for Chapter 6—Solvable and Nilpotent Groups

6.1. Take G as the symmetric group S_3, and $N = \langle(123)\rangle$.

6.3. By induction on i. The result is trivial for $i = 0$, so suppose that $i \geqslant 1$. If ${}^{i-1}G \subseteq Z_{k-i+1}$, then ${}^{i}G = [G, {}^{i-1}G] \subseteq [G, Z_{k-i+1}] \subseteq Z_{k-i}$, because Z_{k-i+1}/Z_{k-i} is the center of G/Z_{k-i}.

6.4. If ${}^{k}G = 1$, we show by induction that ${}^{k-i}G \subseteq Z_i$. The result holds for $i = 0$. Also, if ${}^{k-i+1}G \subseteq Z_{i-1}$, then $[{}^{k-i}G, G] = {}^{k-i+1}G \subseteq Z_{i-1}$, and so ${}^{k-i}GZ_{i-1}/Z_{i-1}$ lies in the center Z_i/Z_{i-1} of G/Z_{i-1}. Therefore ${}^{k-i}G \subseteq Z_i$.
In particular, ${}^{0}G = G \subseteq Z_k$, and the result follows from Problem 6.3.

6.5. The center Z of a finite p-group is not trivial, by Problem 2.53. Hence, by induction on the order of G, G/Z has an upper central series

$$1 \subset Z_1/Z \subset \cdots \subset Z_k/Z = G/Z.$$

Then

$$1 \subset Z \subset Z_1 \subset \cdots \subset Z_k = G$$

is the upper central series for G, and so G is nilpotent.

6.6. It is readily seen that in this case G is a direct product of its Sylow subgroups. The result then follows from 6.T.3 and Problem 6.5.

6.7. Define a sequence of subgroups of N as follows:

$$N_0 = N \quad \text{and} \quad N_i = [N_{i-1}, K] \quad \text{for } i = 1, 2, \ldots$$

By straightforward induction on i, we find

(a) $N_i \subseteq {}^{i-1}K$ for $i = 1, 2, \ldots$;
(b) $N_0 \supseteq N_1 \supseteq N_2 \supseteq \cdots$ is a chain of normal subgroups of G.

Since K is nilpotent, we conclude from (a) that, for some integer $s \geqslant 0$, $N_s \neq 1$ and $N_{s+1} = 1$. Since N is a minimal normal subgroup of G, (b) implies that $N_s = N$. Hence $1 = N_{s+1} = [N,K]$ as required.

6.8. The chain

$$G = {}^0GH \supseteq {}^1GH \supseteq \cdots \supseteq {}^kGH = H$$

is a normal series of length k from G to H.

6.9. A maximal subgroup H of a group G which is also subnormal in G is clearly normal in G. Hence the first part follows from Problem 6.8.

Conversely, if G is a finite group, and each maximal subgroup of G is normal in G, then we shall show that G is nilpotent. Let us suppose the contrary. Then, by Problem 6.6, some Sylow subgroup P of G is not normal in G. Let M be a maximal subgroup of G containing $N(P;G)$. Let x be an element of G not in M. Then $x^{-1}Px \subseteq x^{-1}Mx = M$, so that $x^{-1}Px$ is a Sylow p-group of M. Hence, for some $u \in M$, $x^{-1}Px = u^{-1}Pu$, by Problem 2.23. Hence $xu^{-1} \in N(P;G) \subseteq M$, and so $x \in M$. This contradicts the choice of x. Thus every Sylow subgroup of G is normal in G, and the result follows.

6.10. From Problems 6.8 and 4.21.

6.11. Let N be a minimal normal subgroup of G. The derived group N' is a proper subgroup of N, and N' is a normal subgroup of G by Problem 3.20. Hence, by the choice of N, $N' = 1$; that is, N is abelian. The only simple groups which are subgroups of N must have prime order. The result now follows from Problem 4.12.

6.12. Let N be a minimal normal subgroup of G. If $N \subseteq M$, then, by induction on the order of G, we may conclude that $|G/N{:}M/N| = |G{:}M|$ is a prime power. On the other hand, if N does not lie in M, then $G = MN$, because M is maximal in G. Hence $|G:M| = |MN:M| = |N:N \cap M|$, and the result follows from Problem 6.11.

6.13. We suppose that G has the upper central series (6.2). Each of the subgroups $Z_i \cap N$ $(i = 0, 1, \ldots, k)$ is normal in G. For some $i > 0$, $Z_i \cap N \neq 1$ and $Z_{i-1} \cap N = 1$. Then $[G, Z_i \cap N] \subseteq Z_{i-1} \cap N = 1$, and so $Z_i \cap N$ is a nontrivial subgroup of $Z(G)$. This proves the first part. Finally, if G is finite and N is a minimal normal subgroup of G, then $Z_i \cap N$ must equal N, and so $N \subseteq Z(G)$. Since each subgroup of $Z(G)$ is normal in G, N must be simple; that is, N is of prime order.

6.14. Let P be the set of all p-elements of G. It is sufficient to show that, for all $x,y \in P$, $xy \in P$. However, the subgroups $\langle x\rangle = X$ and $\langle y\rangle = Y$ are finite subnormal subgroups of G, by Problem 6.8, and so $H = \langle X,Y\rangle$ is a finite subgroup of G, by Problem 4.22. Since H is nilpotent, the p-elements x and y lie in the unique Sylow p-group of H (see Problem 6.10), and therefore xy is a p-element as required (see [57]).

6.15. The subgroup $Z = Z(N)$ is characteristic in N, and so is normal in G, by Problem 3.20. Since G/N is cyclic, $G = \langle N,u\rangle$ for some element u in G. Let $H = \langle Z,u\rangle = Z\langle u\rangle$. Since $\langle u\rangle$ is subnormal in H, there exists $z \neq 1$ in Z such that $z^{-1}\langle u\rangle z = \langle u\rangle$. Then either $z \in Z(G)$ or $[u,z] \neq 1$. In the latter case $[u,z] \in Z \cap \langle u\rangle$, and so $[u,z] \in Z(G)$. Hence, in either case, $Z(G) \neq 1$.

6.16. Define the ascending chain of subgroups, $H_0, H_1, \ldots,$ in G by

$$H_0 = H \qquad \text{and} \qquad H_i = N(H_{i-1}; G) \neq H_{i-1} \qquad (i = 1, 2, \ldots).$$

Since this is a proper ascending chain, it follows from Problem 1.34 that it is finite. Hence, for some integer m, $H_m = G$. Then

$$G = H_m \supset H_{m-1} \supset \cdots \supset H_0 = H$$

is a normal series from G to H.

6.17. Choose $a \neq 1$ in G. We now choose a sequence of elements, $a_0, a_1, \ldots,$ in G, and a sequence of subgroups, $A_0, A_1, \ldots,$ of G as follows. Put $a_0 = a$ and $A_0 = \langle a\rangle$. For $i \geqslant 1$, choose a_i to lie in $N(A_{i-1}; G)$ but not in A_{i-1}, and put $A_i = \langle A_{i-1}, a_i\rangle$. Then A_{i-1} is normal in A_i, and the factor group A_i/A_{i-1} is cyclic $(i = 1, 2, \ldots)$. Since $Z(A_0) \neq 1$, it follows inductively, using Problem 6.15, that $Z(A_i) \neq 1$ for each i. Finally, since $A_0 \subset A_1 \subset \cdots$ is a proper ascending chain, it follows from Problem 1.34 that $G = A_n$ for some integer n. Hence $Z(G) = Z(A_n) \neq 1$.

6.18. Since the given conditions are also satisfied by the groups G/Z_i $(i = 1, 2, \ldots)$, we see, from Problem 6.17, that the upper central series (6.2) of G is a proper ascending chain of subgroups in G. Therefore the result follows from Problems 1.34 and 6.4. (For related results see [2], Section 63. The results of Problems 6.15 to 6.18 appear in [58].)

6.19. We proceed by induction on i. The result holds for $i = 1$ for all $j \geqslant i$. If $i \geqslant 2$, and $[Z_j, {}^{i-2}G] \subseteq Z_{j-i+1}$ for all $j \geqslant i$, then

$$[Z_j, {}^{i-1}G] = [Z_j, [{}^{i-2}G, G]] \subseteq [[Z_j, G], {}^{i-2}G][[Z_j, {}^{i-2}G], G]$$
$$\subseteq [Z_{j-1}, {}^{i-2}G][Z_{j-i+1}, G] \subseteq Z_{j-i},$$

using Problem 5.3(a).

6.20. Since ${}^{k-i}G \subseteq Z_i$, by Problem 6.3, the result follows from Problem 6.19.

6.21. Since $k - n - 1 \leqslant n$, therefore $[{}^nG, {}^nG] \subseteq [{}^nG, {}^{k-n-1}G] = 1$, by Problem 6.20.

6.22. By Problem 6.21, ${}^nG$ is an abelian group if n is the greatest integer not exceeding $\frac{1}{2}k$. Since $G/{}^nG$ has class n and solvable length at least $l - 1$, it follows by induction on the class that $2^{l-2} \leqslant n \leqslant k/2$. Hence $2^{l-1} \leqslant k$. (For the results of Problems 6.19 to 6.21, see [59] and [9], page 158.)

6.23. Since each H_i is a normal subgroup of G, by Problem 5.4, therefore each U_n is a normal subgroup of G. Using Problem 5.3, we see that

$$[U_n, G] = \prod_{i=0}^{n} [[H_i, H_{n-i}], G] \subseteq \prod_{i=0}^{n} [[H_i, G], H_{n-i}][H_i, [H_{n-i}, G]]$$
$$= \prod_{i=0}^{n} [H_{i+1}, H_{n-i}][H_i, H_{n-i+1}] \subseteq U_{n+1}.$$

6.24. Define the chain of subgroups, $U_0, U_1, \ldots$, in G as in Problem 6.23. If we use Problem 6.23 and note that ${}^mG \subseteq U_0 = H'$, it follows inductively that $H_{m+2i-1} \subseteq {}^{m+2i-1}G \subseteq U_{2i-1} \subseteq [H, H_i]$ for $i = 1, 2, \ldots$ Hence, if $H_i \subseteq {}^iG \subseteq {}^jH$, then $H_{m+2i-1} \subseteq {}^{m+2i-1}G \subseteq {}^{j+1}H$. Since $H_m \subseteq {}^mG \subseteq {}^1H$, we therefore have (by induction on j) that $H_i \subseteq {}^iG \subseteq {}^jH$ for $i = (2^j - 1)m - 2^{j-1} + 1$. The result then follows. (A more precise bound to the class of G is given in [60]. See also [4], Section 6.6.)

6.25. Using Problem 5.3(b), we have that

$$[G, {}^{n-1}H] = [H, {}^{n-1}H][K, {}^{n-1}H] \subseteq {}^nH({}^{n-1}H \cap {}^0K),$$

and
$$[G,{}^{n-1}K] \subseteq {}^{n}K({}^{0}H \cap {}^{n-1}K).$$

Also, we have
$$[G,{}^{i}H \cap {}^{j}K] = [H,{}^{i}H \cap {}^{j}K][K,{}^{i}H \cap {}^{j}K] \subseteq ({}^{i+1}H \cap {}^{j}K)({}^{i}H \cap {}^{j+1}K).$$

The result now follows by induction on n, starting from ${}^{0}G = {}^{0}H{}^{0}K$ (see [61]).

6.26. Since either $i \geqslant h$ or $h + k - i - 1 \geqslant k$, it follows that ${}^{i}H \cap {}^{h+k-i-1}K = 1$ for $i = 0, 1, \ldots, h + k - 1$. Hence, from Problem 6.25, ${}^{h+k}G = 1$.

6.27. Let us suppose that x is an element of G normalizing N. If $x \notin N$, then H and $x^{-1}Hx$ are distinct normal nilpotent subgroups of N. Hence, by Problem 6.26, $Hx^{-1}Hx$ is a nilpotent subgroup of G properly containing H. This contradicts the choice of H (see [61]).

6.28. Clearly, $R(G)$ is a characteristic subset of G. To show that it is a subgroup of G we must show that, if x and y lie in $R(G)$, then x^{-1} and xy also lie in $R(G)$. But suppose that M and N are normal nilpotent subgroups of G containing x and y, respectively. Then $x^{-1} \in M$ and $xy \in MN$, and so the result follows from Problem 6.26.

It remains to show that, if $x_1, x_2, \ldots, x_n$ lie in $R(G)$ and generate a subgroup H of G, then H is nilpotent. In fact, if N_i is a normal nilpotent subgroup of G containing x_i ($i = 1, 2, \ldots, n$), then, using Problem 6.26, we find that $N_1N_2 \cdots N_n$ is a normal nilpotent subgroup of G containing H. Hence H is nilpotent.

6.29. Let $x \in R(G)$ and $y \in G$. Since $R(G)$ is normal in G, therefore $[x,y] \in R(G)$. Hence, by Problem 6.28, $H = \langle x,[x,y]\rangle$ is nilpotent of class k, say. By induction on i, we find that ${}_{i+1}[x,y] \in {}^{i}H$ for $i = 1, 2, \ldots,$ and so ${}_{k+1}[x,y] = 1$.

6.30. Because of Problem 6.29, it is sufficient to show that each nilelement x of G lies in $R(G)$; that is, x is contained in a normal nilpotent subgroup of G. Let M be a minimal normal subgroup of G. By induction on the order of G, we may suppose that the nilelement Mx of G/M lies in a normal nilpotent subgroup N/M of G/M. Let N be chosen as small as possible, and consider the two cases $N \neq G$ and $N = G$.

If $N \neq G$, then, by induction on the order of the group, $x \in R(N)$. Since $R(N)$ is a normal nilpotent subgroup of G, using Problems 6.28 and 3.20, therefore $x \in R(G)$ as required.

If $N = G$, then it follows that $G = \langle M,x \rangle$. (Otherwise, some maximal subgroup H/M of G/M contains Mx, and H/M is normal in G/M, by Problem 6.9, which contradicts the choice of N.) Next, we note that $C = C(x;M)$ is a subgroup of the abelian group M (see Problem 6.11) which is normalized by x. Hence C is a normal subgroup of G contained in M. Now $C \neq 1$ because if $u \neq 1$ lies in M, then for some integer $i \geqslant 0$, the element ${}_i[x,u] = v$, say, is nontrivial, while $[x,v] = {}_{i+1}[x,u] = 1$. (Compare with the proof of Problem 5.18.) Therefore, by the choice of M, $C = M$. Thus $x \in Z(G)$ and $G = Z(G)M$. In particular, G is nilpotent of class at most 2 and the result follows. (For the results of Problems 6.28 to 6.30 see [46].)

6.31. Let us consider the group $G/{}^2G$. Then ${}^1G/{}^2G \subseteq Z(G/{}^2G)$, and $G/{}^2G/{}^1G/{}^2G$ is isomorphic to $G/{}^1G$ and is therefore cyclic. Hence, by Problem 1.15, $G/{}^2G$ is abelian, and this implies ${}^2G = {}^1G$ by 5.T.1. It then follows by induction that ${}^iG = {}^1G$ for $i = 1, 2, \ldots$

6.32. Since H' is a normal cyclic subgroup of G, by Problem 3.20, therefore $G/C(H';G)$ is abelian, by Problem 3.18. Hence, from 5.T.1, $H \subseteq G' \subseteq C(H';G)$, and so $H' \subseteq Z(H)$. Thus ${}^2H = 1$, and so $H' = 1$, by Problem 6.31.

The second part follows immediately now by considering $G/G^{(k+1)}$.

6.33. By Problem 6.13, there is a coset $zM \neq M$ lying in $N/M \cap Z(G/M)$. Then $K = \langle z,M \rangle$ has the required properties.

6.34. By Problem 6.33, we may choose successively normal subgroups G_i of G such that $G_0 = 1, G_1, \ldots,$ is a proper ascending chain and G_{i+1}/G_i is cyclic for each $i \geqslant 0$. Because G satisfies the maximal condition on normal subgroups, this chain must have a maximal element G_n, say, and clearly $G_n = G$. The assertion now follows (see also [62]).

6.35. Let us consider the normal series for G defined as in Problem 6.34. Then, for each subgroup H of G,

$$G_0 \cap H = 1 \subseteq G_1 \cap H \subseteq \cdots \subseteq G_n \cap H = H$$

is a normal series for H. Since $G_{i-1} \cap H = G_{i-1} \cap (G_i \cap H)$, therefore $(G_i \cap H)/(G_{i-1} \cap H) \simeq G_{i-1}(G_i \cap H)/G_{i-1} \subseteq G_i/G_{i-1}$ for each i. Thus the factor groups in the given normal series for H are all cyclic. We choose $x_i \in H$ so that $G_i \cap H = \langle G_{i-1} \cap H, x_i \rangle$ for $i = 1, \ldots, n$. Then $H = \langle x_1, \ldots, x_n \rangle$. (Compare with [2], Sections 63 and 59.)

6.36. We suppose that $A \subset C(A;G)$. Then by Problems 1.14 and 6.33 there exists a normal subgroup K of G such that $A \subset K \subseteq C(A;G)$ and

K/A is cyclic. Then $A \subseteq Z(K)$, and so K is abelian, by Problem 1.15. This is contrary to the choice of A.

6.37. We proceed by induction on n, noting that the result is true for $n = 0$. If $n \geqslant 1$, then $M = N \cap Z(G) \neq 1$, by Problem 6.33, and N/M is a normal subgroup of order p^m $(m < n)$ in G/M. The result now follows by applying the induction hypothesis.

6.38. Let us suppose that A is not cyclic. Then, by Problem 3.4, A has at least two subgroups of order p. Therefore we can choose $a \in A$ such that $a^p = 1$ and $a \notin N$. Then, for each $x \in G$, $[x,a] \in N$. But $[x,a]^p = [x,a^p] = 1$ because A is abelian. Therefore, since N has only one subgroup of order p, say P, $[x,a] \in P$ for all $x \in G$. Hence each conjugate $x^{-1}ax$ lies in aP, and so a has at most p conjugates.

6.39. Put $N = Z(G')$ and suppose that, contrary to the assertion, $N \neq G'$. Then, by Problem 6.33, there exists a subgroup K which is normal in G such that $N \subset K \subseteq G'$ and K/N is cyclic. Let A/N be the (unique) subgroup of order p in K/N. Then A is normal in G, by Problem 3.20. Since $N \subseteq Z(A)$ and A/N is cyclic, A is abelian by Problem 1.15. If A is cyclic, then $C(A;G)$ contains G', from Problem 3.18 and 5.T.1, and so $A \subseteq Z(G')$ contrary to hypothesis. Thus A is not cyclic, and so, by Problem 6.38, $A = \langle N,a \rangle$, where $|\, G:C(a;G) \,| \leqslant p$. Thus $C(a;G)$ is a maximal subgroup of the nilpotent group G, and so is normal in G, by Problem 6.9. Therefore $G' \subseteq C(a;G)$, and so $A = \langle N,a \rangle \subseteq Z(G')$, contrary to the choice of A. Thus $N = G'$ as asserted.

6.40. The number of subgroups of index p in H is congruent to 1 (mod p), by Problem 2.22. Let the set of these subgroups be $\Omega = \{K_0, K_1, \ldots\}$, and define $K_i^x = x^{-1}K_ix$ $(x \in G;\, K_i, x^{-1}K_ix \in \Omega)$. Then G operates as a permutation group on Ω. Since G is a p-group, the lengths of the orbits in Ω are all powers of p, by Problem 2.12, and so the number of orbits of length 1 is congruent to 1 (mod p). If $\{K\}$ is such an orbit, then K is the required normal subgroup of G.

6.41. Let us suppose that, on the contrary, $G'' \neq 1$. Then, by Problem 6.40, there exists a normal subgroup K of G contained as a subgroup of index p in G''. If we consider the group G/K, then $(G/K)' = G'/K$ and $(G/K)'' = G''/K \neq 1$, using 5.T.3. Therefore, by Problem 6.39, the center Z/K of G'/K is not cyclic, and so $|\, Z:K \,| \geqslant p^2$. Hence $|\, G':Z \,| \leqslant p$, and so $G'/K/Z/K$ is cyclic. Therefore G'/K is abelian, by Problem 1.15. This contradicts the choice of K. (The results of Problems 6.39 and 6.41 are due to Burnside. See [63].)

6.42. If G is nilpotent, take $K = G$. Otherwise, some maximal subgroup M of G is not normal in G, by Problem 6.9. We may suppose, by induction on the order of the group, that there exists a nilpotent subgroup K of M such that $K^M = M$. Then $K^G \supseteq M$, and K^G is normal in G. Therefore $K^G = G$ (see [64]).

6.43. Let P be a Sylow p-group of G. Then P is normal in G by Problem 6.10, and so the result follows from Problem 2.49.

6.44. Let G_α be a stabilizer of G, and let M be a maximal subgroup of G containing G_α. By Problem 6.9, M is normal in G and so is of index p in G. Since $|M:G_\alpha| = n/p$, it is readily seen that α^M is an orbit of length n/p for M. From Problem 2.43, we know that M has p orbits $\Omega_1 = \alpha^M$, $\Omega_2, \ldots, \Omega_p$. If we use Problem 2.36, it follows that $M|\Omega_i$ has the same solvable length as M, for each i. Since M is maximal in G, the solvable length of M is at least $l - 1$. Hence we may suppose (by induction on n) that p^{l-1} divides $|\Omega_i| = n/p$. Thus $p^l \mid n$.

6.45. If p divides $|G|$, then $p \mid n$, by Problem 6.43. Let G_α be a stabilizer of G. Since $|G:G_\alpha| = n$, by Problem 2.30, and G is a direct product of its Sylow subgroups, by Problem 6.10, it is readily shown that G has a normal subgroup M which contains G_α and has index p in G. As in the proof of Problem 6.44, M has p orbits $\Omega_1, \Omega_2, \ldots, \Omega_p$, each of length n/p. Using Problem 2.36, we see that, if r is the highest exponent to which p divides the order of $M|\Omega_i$, then the corresponding exponent for M is no greater than rp. By induction on n, we may therefore conclude that the highest exponent to which p divides $|G|$ is at most

$$rp + 1 \leqslant \frac{p^{k-1} - 1}{p - 1} \cdot p + 1 = \frac{p^k - 1}{p - 1}.$$

The result then follows.

6.46. For any prime p,

$$\log p^{(p^k-1)/(p-1)} = (p^k - 1)\frac{\log p}{p - 1} \leqslant (p^k - 1)\log 2.$$

Hence the product given in Problem 6.45 is less than or equal to

$$\prod_{p|n} 2^{p^k-1} \leqslant 2^{n-1}.$$

This proves the case when G is transitive. In general, if G has orbits $\Omega_1, \ldots, \Omega_k$ of lengths $n_1, \ldots, n_k$, respectively, then $| G|\Omega_i | \leqslant 2^{n_i-1}$ from above. Hence, by Problem 2.36,

$$| G | \leqslant \prod_{i=1}^{k} 2^{n_i-1} \leqslant 2^{n-1}.$$

[The bound is actually attained when G is a Sylow 2-group of S_{2^k} $(k = 1, 2, \ldots)$.]

6.48. Let us suppose that G has at least two abelian subgroups A and B of index p. Since A and B are normal in G, by Problem 6.9, $G = AB$. Since A and B both centralize $A \cap B$, therefore $A \cap B \subseteq Z(AB) = Z(G)$. But $| G:A \cap B | = p^2$, by 1.T.2, and $G/Z(G)$ is not cyclic, by Problem 1.15, and so $A \cap B = Z(G)$. Thus $| G:Z(G) | = p^2$. Every abelian subgroup C of index p in G must contain $Z(G)$, because otherwise G would be abelian. Therefore G has as many abelian subgroups of index p as $G/Z(G)$ has. Since $G/Z(G)$ is noncyclic, by Problem 1.15, and has order p^2, this number is $p + 1$.

6.49. Let $Z = Z(G)$. Since G has class 3, G/Z is not abelian, and so it has order 8 and $| Z | = 2$. By Problem 1.16, not every element of G/Z has order 2, so let xZ be an element of order 4 in G/Z. Then $A = \langle x,Z \rangle$ is abelian, and, by Problem 6.48, A is the only abelian subgroup of index 2 in G. We finally prove that A is cyclic. If A were not cyclic, then $x^4 = 1$, and so $A = \langle x \rangle \times Z$. Then $B = \{u^2 \mid u \in A\} = \langle x^2 \rangle$ is a characteristic subgroup of A, and so B is normal in G, by Problem 3.20. But then $B \cap Z = 1$ contradicts Problem 6.13. Thus we conclude that A is cyclic.

***6.50.** Any nonabelian group of order 8 has class 2, by Problem 6.31, and so we shall consider the case $n \geqslant 4$. If we use the technique of Problem 6.49 and induction on n, it is straightforward to show that any group G with the desired property has exactly one cyclic subgroup of index 2. An examination of the possible types of groups of order 2^n with a cyclic subgroup of index 2 (see [18], Section 109 or [1], Theorem 12.5.1) then shows that G must be isomorphic to one of the following: a generalized quaternion group (see Problem 5.29); a dihedral group (see Problem 1.6); or a group generated by two elements x and y with the sole relations $x^{2^{n-1}} = y^2 = 1$ and $y^{-1}xy = x^{-1+2^{n-2}}$.

***6.51.** See [18], Section 118.

***6.52.** Let P be a Sylow p-group of G, and then consider the subgroup $D = P \cap x^{-1}Px$, where x is chosen to make D as large as possible. See [55], page 185, or [9], page 138.

***6.53.** Let $D = M \cap L$, where M and L are distinct maximal subgroups of G, and choose M and L so that D has largest possible order. Show that D is then normal in G and use Problem 1.42. (See [65] or [5], page 148. A generalization due to B. Huppert is proved in [4], page 237.)

***6.54.** See [1], Section 18.2, or [4], page 208.

Solutions for Chapter 7—The Group Ring and Monomial Representations

7.1. For any $x,y \in G$, we have

$$\theta^G(x)\theta^G(y) = [\dot{\theta}(r_i^{-1}xr_j)][\dot{\theta}(r_i^{-1}yr_j)] = [\sum_{k=1}^{n} \dot{\theta}(r_i^{-1}xr_k)\dot{\theta}(r_k^{-1}yr_j)].$$

But $\dot{\theta}(r_i^{-1}xr_k)\dot{\theta}(r_k^{-1}yr_j)$ is nonzero only if we have both $r_i^{-1}xr_k$ and $r_k^{-1}yr_j$ lying in H, that is, both $x^{-1}r_i$ and yr_j lying in the same coset r_kH. For given i and j there is at most one such k. There is such a k exactly when $(x^{-1}r_i)^{-1}(yr_j) = r_i^{-1}xyr_j$ lies in H. Thus

$$\sum_{k=1}^{n} \dot{\theta}(r_i^{-1}xr_k)\dot{\theta}(r_k^{-1}yr_j) = \dot{\theta}(r_i^{-1}xyr_j).$$

Hence $\theta^G(x)\theta^G(y) = \theta^G(xy)$ for all $x,y \in G$, and so θ^G is a homomorphism. Finally, $\theta^G(x) = \text{diag}\,(1, 1, \ldots, 1)$ if and only if $r_i^{-1}xr_i \in N$ for each i. Thus the kernel of θ^G is

$$\bigcap_{i=1}^{n} r_iNr_i^{-1} = \bigcap_{x\in G} x^{-1}Nx.$$

[*Note:* There is a close relation between the monomial representations and the permutation representation given in Problem 2.17. The latter is equivalent to a special case of the former.]

7.2. Choose $\mathbf{u} = [\dot{\theta}(r_i^{-1}s_j)]$. Then $\mathbf{u}$ is an $n \times n$ monomial matrix since $\dot{\theta}(r_i^{-1}s_j)$ is nonzero exactly when the left coset representatives r_i and s_j correspond to the same left coset of H. In particular, $\mathbf{u}$ has an inverse, $\mathbf{u}^{-1}$.

Now, for any $x \in G$, we have

$$\theta^G(x)\mathbf{u} = [\sum_{k=1}^{n} \dot{\theta}(r_i^{-1}xr_k)\dot{\theta}(r_k^{-1}s_j)].$$

Since $\dot{\theta}(r_i^{-1}xr_k)\dot{\theta}(r_k^{-1}s_j)$ is nonzero if and only if $x^{-1}r_i$ and s_j both lie in the same coset r_kH, we find, as in the solution of Problem 7.1, that

$$\theta^G(x)\mathbf{u} = [\dot{\theta}(r_i^{-1}xs_j)].$$

Similarly, we find that $\mathbf{u}\theta_1^G(x) = [\dot{\theta}(r_i^{-1}xs_j)]$, and the result follows.

7.3. Since S is abelian, the concept of determinant is applicable to monomial matrices over S. Also, we have

$$\begin{aligned}\tau(x)\tau(y) &= (\det \theta^G(x))(\det \theta^G(y)) = \det \theta^G(x)\theta^G(y)\\ &= \det \theta^G(xy) = \tau(xy).\end{aligned}$$

Thus τ is a homomorphism of G into S.

From Problem 7.2, we have

$$\det \theta_1^G(x) = (\det \mathbf{u}^{-1})(\det \theta^G(x))(\det \mathbf{u}) = \det \theta^G(x),$$

and so the value of the transfer is independent of the particular set of left coset representatives used in defining the monomial representation.

7.4. The subgroup $N = K \cap A$ is normalized by both A and K, and so it is a normal subgroup of $G = KA$. The group G/N contains two Hall subgroups K/N and HN/N of order n, using 1.T.5, and an abelian normal subgroup A/N of index n. Therefore, by Problem 3.22, there is $x \in G$ such that $x^{-1}Hx \subseteq x^{-1}HNx = K$.

7.5. Let θ be the identity homomorphism of A onto itself; that is, $\theta(a) = a$ for all $a \in A$. Writing $G^* = \theta^G(G)$ and $A^* = \theta^G(A)$, we have $G^* \simeq G$ and $A^* \simeq A$, by Problem 7.1. Now $P = S(n) \cap G^*D(n,A)$ is a group of permutation matrices, and

$$\begin{aligned}PD(n,A) &= \{S(n) \cap G^*D(n,A)\}D(n,A) = S(n)D(n,A) \cap G^*D(n,A)\\ &= G^*D(n,A),\end{aligned}$$

using Problem 1.38 and 7.T.4. Therefore

$$| P | = | P:P \cap D(n,A) | = | PD(n,A):D(n,A) | = | G^*:G^* \cap D(n,A) |$$
$$= | G^*:A^* | = n.$$

Thus P is a Hall subgroup of $PD(n,A)$. Since $D(n,A)$ is abelian and of order $| A |^n$, and n divides $| G^* |$, it follows from Problem 7.4 that G^* contains a subgroup of order n conjugate to P in $G^*D(n,A)$. Thus G, which is isomorphic to G^*, has a subgroup of order n as asserted. (For other proofs of this result see [1], page 223; [66]; or [4], page 144.)

7.6. We proceed by induction on $| G |$. Consider two cases:

(a) *H contains a subgroup $N \neq H$ or 1 with N normal in G.* Then, by the induction hypothesis, there exists a complement L/N for H/N in G/N. But N is a Hall subgroup in L, so we can apply induction again to find a complement K of N in L. It is readily shown that K is a complement for H in G.

(b) *H contains no subgroup $N \neq H$ or 1 with N normal in G.* If H is a p-group, then this implies that $H' = 1$ because H' is normal in G, by Problem 3.14. Then H is abelian, and the result follows from Problem 7.5. If H is not of prime power order, then let P be a nontrivial Sylow subgroup of H. By Problem 3.13, $G = N(P;G)H$, and, since P is not normal in G, $N(P;G)$ is a proper subgroup of G. Hence, by the induction hypothesis, $N(P;G) \cap H$ has a complement K in $N(P;G)$. It is readily verified that K is a complement for H in G. (This theorem is due to Schur. See [1], page 224.)

7.7. We proceed by induction on $| G |$. (a) *Suppose that H is solvable.* If H is abelian, then the result is proved in Problem 3.22. In the other case, H' is a nontrivial normal subgroup of G, by Problem 3.20. The factor group G/H' has a normal Hall subgroup H/H' with two complements KH'/H' and LH'/H' in G/H'. By the induction hypothesis, $x^{-1}KH'x/H' = LH'/H'$ for some $x \in G$. Then $LH' = x^{-1}KxH'$, which is a proper subgroup of G, and the normal Hall subgroup H' has two complements L and $x^{-1}Kx$ in LH'. Hence, by the induction hypothesis, L is conjugate to $x^{-1}Kx$, and so L is conjugate to K.

(b) *Suppose that G/H is solvable; that is, K and L are solvable.* Let M be a maximal subgroup of K containing K'. Since K/K' is abelian, M is a normal subgroup of index p, say, in K for some prime p. Put $M_1 = L \cap MH$. Then M and M_1 are both complements of the normal Hall subgroup H in MH. Thus, by the induction hypothesis, $x^{-1}M_1x = M$ for some $x \in MH$. Then $M \subseteq x^{-1}Lx \cap K$, and M is normal in both $x^{-1}Lx$ and K. Thus M is a normal subgroup of $S = \langle x^{-1}Lx,K \rangle$. Finally, let P

be a Sylow p-group of K. Since P is a Sylow p-group of S, there exists $y \in S$ such that $y^{-1}Py$ is a Sylow p-group of $x^{-1}Lx$. Since M is normal in S, $K = MP \subseteq yx^{-1}Lxy^{-1}$. Thus K is conjugate to L. (See [9], pages 162 and 163. For related results see [67].)

7.8. Let $r_1, r_2, \ldots, r_h$ be a set of left coset representatives of $Z = Z(G)$ in G. Let θ be the identity mapping of Z onto itself. Then we consider the monomial representation θ^G of G corresponding to the given set of coset representatives, and the transfer τ of G into Z.

For each $x \in G$ we define the mapping $\pi\colon i \to j$ of the set $\{1, 2, \ldots, h\}$ into itself by the condition that j is the unique integer such that $r_i^{-1}xr_j \in Z$. Clearly, π is a permutation and

$$\tau(x) = (r_1^{-1}xr_{\pi(1)})(r_2^{-1}xr_{\pi(2)}) \cdots (r_h^{-1}xr_{\pi(h)}).$$

Let $(i_1i_2 \cdots i_s)$ be a cycle of π (see 2.T.1). Since $\tau(x)$ is a product of commuting elements, we may collect together the terms corresponding to this cycle. Their contribution is

$$(r_{i_1}^{-1}xr_{i_2})(r_{i_2}^{-1}xr_{i_3}) \cdots (r_{i_s}^{-1}xr_{i_1}) = r_{i_1}^{-1}x^sr_{i_1} = x^s,$$

since the product lies in Z. Thus the product of the contributions of all the cycles of π is $\tau(x) = x^h$. The result now follows from Problem 7.3.

7.9. The homomorphism of G into Z defined in Problem 7.8 has an abelian image. This implies that G' is contained in the kernel of the homomorphism, by 5.T.1. Hence, for each $x \in G'$, $x^h = 1$. (For the results of Problems 7.8 and 7.9, see [68].)

7.10. (a) *Suppose that the condition* (*) *holds.* Then from each double coset HxH choose a particular y such that $C(y;H) = y^{-1}Hy \cap H$. Let $C_y = \{u^{-1}yu \mid u \in H\}$. Each left coset of H in $HxH = HyH$ has the form $u^{-1}xH = u^{-1}xuH$ for some $u \in H$, and so C_y contains a complete set of left coset representatives for H in HxH. But $|C_y| = |H{:}C(y;H)| = |H{:}H \cap y^{-1}Hy|$, which is the number of left cosets of H in HxH, by Problem 2.13. Thus C_y consists of exactly one representative for each left coset of H in HxH. The union of all the sets C_y (one for each double coset) gives an exceptional set of left coset representatives for H in G.

(b) *Suppose that R is an exceptional set of left coset representatives for H in G.* For any $x \in G$, we choose $y \in HxH \cap R$ and assert that $y^{-1}Hy \cap H = C(y;H)$. Clearly, $C(y;H) \subseteq y^{-1}Hy \cap H$. On the other hand, if $u \in y^{-1}Hy \cap H$, then $u = y^{-1}vy$ for some $v \in H$. Then $vyv^{-1}H = vyH = yuH = yH$. Since R is an exceptional set of left coset representatives, this implies that $vyv^{-1} = y$, and so $u = v \in C(y;H)$. Thus $y^{-1}Hy \cap H \subseteq C(y;H)$. The result now follows.

7.11. Let $r_1, r_2, \ldots, r_n$ be an exceptional set of coset representatives for H in G. Let θ be the identity mapping of H onto itself, and consider the transfer τ of G into H. Because we are dealing with an exceptional set of coset representatives, the condition that x and $r_i^{-1}xr_j$ both lie in H implies that $r_i^{-1}xr_j = x$. Thus in the monomial representation θ^G corresponding to this set of coset representatives, each matrix $\theta^G(x)$ $(x \in H)$ has all its nonzero entries equal to x. Therefore $\tau(x) = x^n$ for all $x \in H$. Since $n = | G:H |$ is relatively prime to $| H |$, it follows from Problem 3.21 that $\{x^n \mid x \in H\} = H$. Thus the transfer τ maps G *onto* H. Let K be the kernel of τ. Then $| G:K | = | H |$, and so, from Problem 1.8, K is the required normal complement.

Conversely, if H has a normal complement K in G, then K is an exceptional set of coset representatives. (See [69]. The hypothesis that H be abelian has been considerably relaxed in a series of papers by Kochendörffer and Zappa. A final solution, using character theory, was given in [70].)

7.12. We shall show that $x^{-1}Px \cap P = C(x;P)$ for all $x \in G$. The result will then follow from Problems 7.10 and 7.11. Let us suppose that $u \in x^{-1}Px \cap P$. Then $u = x^{-1}vx$ for some $v \in P$, and, by Problem 3.25, this implies that $u = v$ and $u \in C(x;P)$. Thus $x^{-1}Px \cap P \subseteq C(x;P)$. The reverse inequality is trivial, so the result is proved. (Compare with the proof given in [1], page 203.)

7.13. Let us suppose that a Sylow p-group P of G has order p^k. Since P is a cyclic normal subgroup of $N(P;G)$, $| N(P;G):C(P;G) |$ divides $\phi(p^k) = p^{k-1}(p-1)$, by Problem 3.18. Since $| N(P;G):C(P;G) |$ also divides n/p^k, using 1.T.1, and $p^{k-1}(p-1)$ is relatively prime to n/p^k, the index is 1. Thus $N(P;G) = C(P;G)$, and the result follows from Problem 7.12.

7.14. First we show that G is solvable. We proceed by induction on the order n of G. If p is the smallest prime dividing n, then $p - 1$ is relatively prime to n. Therefore, by Problem 7.13, $G = PK$ where P is a Sylow p-group, and K is a normal complement of P in G. Using the induction hypothesis, we find that K is solvable. Since $G/K \simeq P$ is solvable, it follows that G is solvable.

Now we suppose that G has solvable length l. The factor groups $G^{(i)}/G^{(i+1)}$ $(i = 0, 1, \ldots, l-1)$ are abelian groups with cyclic Sylow subgroups. Therefore these factor groups are all cyclic, by 3.T.6. Thus, from Problem 6.32, $G'' = 1$ (see [9] page 175).

7.15. A Sylow 2-group P of G cannot be cyclic, by Problem 7.13. Thus, if $|G|$ is not divisible by 8, then P is a noncyclic group of order 4. By Problem 3.35 this implies that $|\text{Aut } P| = 6$, and so $|N(P;G):C(P;G)| = h$, say, divides 6, by Problem 3.17. Since G is simple, and $2 \nmid h$ because P is a Sylow 2-group, we conclude from Problem 7.12 that $h = 3$. Hence $12 = 3|P|$ divides $|G|$ (see [18], Section 245).

7.16. A simple group of order 300 would have six Sylow 5-groups, by Problem 2.23, and the normalizer H of one of these Sylow 5-groups has index 6 in G. But $300 \nmid 6!$, and so we have a contradiction from Problem 2.18.

A simple group G of order 400 would have sixteen Sylow 5-groups of order 5^2. If P is one of these Sylow 5-groups, then $P = N(P;G) = C(P;G)$, using Problems 2.23 and 2.56. Thus Problem 7.12 gives a contradiction.

A simple group G of order 540 would have either six or thirty-six Sylow 5-groups. The first case is eliminated by Problem 2.18; so the normalizer $N(P;G)$ of a Sylow 5-group P has order 15. Since $|\text{Aut } P| = 4$, by Problem 3.8, it follows from Problem 3.17 that $C(P;G) = N(P;G)$. Thus Problem 7.12 gives a contradiction. (For further examples of this kind see [5], Sections 6.2, 6.5, and 10.2.)

7.17. A Sylow p-group P of G cannot be cyclic, by Problem 7.13. Thus, if $p^3 \nmid |G|$, then P is a noncyclic group of order p^2. By Problem 3.35, $|\text{Aut } P| = (p^2 - 1)(p^2 - p) = 2 \cdot p \cdot (p - 1)^2 \cdot \frac{1}{2}(p + 1)$, and this is not divisible by any prime greater than p. But $|N(P;G):C(P;G)|$ divides $|\text{Aut } P|$, by Problem 3.17, and $|G:P|$; so $N(P;G) = C(P;G)$. Thus G is not simple, by Problem 7.12.

7.18. The only orders which are not directly eliminated by Problems 2.23 and 7.17 are $3^3 \cdot 13$ and $3^3 \cdot 5 \cdot 7$. The first of these cases is eliminated by Problem 7.12, with P as a Sylow 13-group. On the other hand, a simple group of order $3^3 \cdot 5 \cdot 7$ would have seven Sylow 3-groups, by Problem 2.23. Since $3^3 \cdot 5 \cdot 7$ does not divide $7!$, Problem 2.18 shows that there is no simple group of this order either.

7.19. We proceed by induction on $|G|$. Let M be a minimal normal subgroup of G. By Problem 6.11, M is an abelian group of order p^k, say, and we consider two cases:

(a) *If* $p \mid m$. Then, by the induction hypothesis, G/M has a subgroup H/M of order m/p^k. Furthermore, since $|KM/M|$ divides m/p^k, there exists $x \in G$ such that $x^{-1}KMx/M \subseteq H/M$. Thus H is a subgroup of order m in G, and $x^{-1}Kx \subseteq H$.

(b) *If* $p \nmid m$. Then, by the induction hypothesis, G/M has a subgroup H_1/M of order m. Furthermore, since $|KM/M|$ divides m, there exists $x \in G$ such that $x^{-1}KMx/M \subseteq H_1/M$. Because M is a normal abelian Hall subgroup of H_1, M has a complement H in H_1, by Problem 7.5, and H is the required subgroup of order m in G. Finally, we show that K is conjugate to a subgroup of H, or equivalently, that $K_1 = x^{-1}Kx$ is conjugate to a subgroup of H. We put $L = K_1M \cap H$, and note that $LM = K_1M \cap HM = K_1M \cap H_1 = K_1M$, by Problem 1.38. Since K_1 is a complement of M in K_1M, $y^{-1}K_1y \subseteq L$ for some $y \in LM$, by Problem 7.4. Since $L \subseteq H$, K_1 is conjugate to a subgroup of H as required. (This result is due to P. Hall. See, for example, [2], Section 60.)

7.20. By Problem 7.19 we may find subgroups C_i of index $p_i^{k_i}$ in G for $i = 1, \ldots, s$. For each i, we define

$$P_i = \bigcap_{j \neq i} C_j.$$

From Problem 1.8, we know that whenever two subgroups A and B have relatively prime indices in G, then $|G:A \cap B| = |G:A|\,|G:B|$. Hence we find $|G:P_i| = \prod_{j \neq i} |G:C_j|$, and so (considering its order) P_i is a Sylow p_i-group of G. Finally, we note that $\bigcap_{k \neq i \text{ or } j} C_k$ is a group of order $|P_i|\,|P_j|$ which contains both P_i and P_j. Hence

$$P_iP_j = P_jP_i = \bigcap_{k \neq i \text{ or } j} C_k$$

by Problem 1.8 (see [71]).

7.21. We proceed by induction on $|G|$. Let M be a minimal normal subgroup of G. By Problem 6.11, M is a p-group for some prime p. Using the induction hypothesis, we can find a self-normalizing nilpotent subgroup L/M of G/M. If P is a Sylow p-group of L, then $M \subseteq P$, and P is normal in L because L/M is nilpotent, by Problem 6.10. By Problem 7.6 there is a complement U of P in L. Now U is a Hall subgroup of L, so $H = N(U;L)$ is self-normalizing in L, by Problem 3.23. Since U is isomorphic to a subgroup of L/M, U is nilpotent. But $H = U \times (P \cap H)$, and so H is also nilpotent. Finally, H is self-normalizing. In fact, since L/M is nilpotent, $L = UMN(UM;L) = MN(U;L)$, by Problem 3.19; that is, $L = MH$. Therefore $N(H;G) \subseteq N(L;G) = L$, and so $N(H;G) = N(H;L) = H$.

7.22. We proceed by induction on $|G|$ and consider two cases.

(a) *A nontrivial normal subgroup N of G lies in $H \cap K$.* Then H/N and K/N are self-normalizing nilpotent subgroups of G/N. Therefore, by the induction hypothesis, $H/N = x^{-1}Kx/N$ for some $x \in G$; that is, $H = x^{-1}Kx$.

(b) *$H \cap K$ contains no nontrivial normal subgroup of G.* Let M be a minimal normal subgroup of G. Then M is a p-group, by Problem 6.11. We shall first show that MH is a self-normalizing subgroup of G. It is sufficient to consider the case $MH \neq G$. Then, for each $x \in N(MH;G)$, $x^{-1}Hx$ is a self-normalizing nilpotent subgroup of MH. Hence, by the induction hypothesis, $x^{-1}Hx = u^{-1}Hu$ for some $u \in MH$; that is, $x \in N(H;G)u = Hu \subseteq MH$. Thus $N(MH;G) = MH$. From this we conclude that MH/M and MK/M are self-normalizing nilpotent subgroups of G/M, and so, by the induction hypothesis applied to G/M, $MH = x^{-1}MKx$ for some $x \in G$. We may suppose that $G = \langle H, y^{-1}Ky \rangle$ for all $y \in G$; otherwise, the result follows immediately from the induction hypothesis. Therefore $G = MH = MK$. Finally, we show that this implies that G is a p-group. Otherwise, for some prime q different from p, there is a nontrivial Sylow q-group Q in H. Since Q is a Sylow q-group of G, some conjugate of Q is equal to a Sylow q-group of K. Thus $Q \subseteq H \cap y^{-1}Ky$ for some $y \in G$. But Q is normal in each of the nilpotent groups H and $y^{-1}Ky$, and so Q is normal in $G = \langle H, y^{-1}Ky \rangle$. Thus Q is a nontrivial normal subgroup of G contained in both H and K, which is contrary to our assumption. Thus we conclude that G is a p-group and $G = H = K$ (see [72]).

7.23. Using Problem 2.22, it is sufficient to show that whenever $P \neq 1$ is a Sylow p-group of H, for some p, then P is a Sylow p-group of G. Let us suppose, on the contrary, that P is not a Sylow p-group of G. Then, by Problem 2.23, P is properly contained in a Sylow p-group P^* of G. Since P is subnormal in P^*, by Problem 6.8, it is properly contained in its normalizer in P^*. Let $x \in N(P;P^*)$ with $x \notin P$. Then $x \notin H$, but $x^{-1}Hx \cap H \supseteq P \neq 1$, contrary to the hypothesis on H.

7.24. If A is normal in G, then $|G:A|$ is prime. Thus $G' \subseteq A$, and so $G'' = 1$. Hence we suppose that A is not normal in G.

First, let us assume that 1 is the only normal subgroup of G in A. For all $x \notin A$, $G = \langle x^{-1}Ax, A \rangle$. Thus $x^{-1}Ax \cap A$ is normal in G and so equals 1. From Problem 7.23 it follows that A is a Hall subgroup of G. Put $|A| = h$. For each prime p dividing h, the Sylow p-group P of A is a Sylow p-group of G. Furthermore, $N(P;G) = A = C(P;G)$, and so P has a normal complement K_p in G, by Problem 7.12. Then

$$K = \bigcap_{p|h} K_p$$

is a normal complement of A in G, and we examine its structure. Let $Q \neq 1$ be some Sylow q-group of K, and put $N = N(Q;G)$. By Problem 3.19, $G = KN$. Since $K \cap N$ is a normal Hall subgroup of N, it has a complement C in N, by Problem 7.6. Since $G = KN = K(K \cap N)C = KC$ and $K \cap C = 1$, both C and A are complements of K in G. Since A is abelian, Problem 7.7 shows that C is conjugate to A, and so C is an abelian maximal subgroup of G. Now Q is normalized by C, and so QC is a subgroup properly containing C. Hence $QC = G$. Thus Q is normal in G, and so $Z(Q)$ is normal in G. Again, since C is maximal, $CZ(Q) = G$. Thus $K = Z(Q)$, and since $G/K \simeq A$, we conclude that $G'' = 1$.

In general, if B is the largest subgroup of A normal in G, then the only normal subgroup of G/B in A/B is 1. Hence it follows from above that $(G/B)'' = 1$. Since B is abelian, $G^{(3)} = 1$. (For alternative proofs, see [73] and [5], page 392. Thompson [74] proves that a group possessing a maximal subgroup which is nilpotent of odd order is solvable. His proof is also given in [4], page 277. For a related result see [75].)

7.25. The alternating group A_5 has no subgroup of order 15 because of 2.T.5 and Problem 2.18.

7.27. Let C be the subgroup of $M(n,B)$ generated by the elements of $D(n,B)$ together with the matrix

$$\mathbf{x} = \begin{bmatrix} 0 & 1 & 0 & \cdots & 0 \\ 0 & 0 & 1 & \cdots & 0 \\ \cdot & & & & \cdot \\ \cdot & & & & \cdot \\ \cdot & & & & \cdot \\ 0 & 0 & 0 & \cdots & 1 \\ 1 & 0 & 0 & \cdots & 0 \end{bmatrix}.$$

Then $|C| = n\,|D(n,B)| = n^{n+1}$. Furthermore, if $\mathbf{y} = \operatorname{diag}(1, b, \ldots, b^{n-1})$ for $b \in B$, then $[\mathbf{x},\mathbf{y}] = \operatorname{diag}(b, b, \ldots, b)$ because $b^n = 1$. Thus C' contains the normal subgroup B_0 of C, where B_0 consists of all matrices of the form $\operatorname{diag}(b, b, \ldots, b)$ $(b \in B)$. Moreover B_0 is isomorphic to B by an obvious mapping.

Solutions for Chapter 8—Frattini Subgroup

8.1. If $\langle x \rangle$ is a cyclic group of order n, then $\Phi(\langle x \rangle) = \langle x^m \rangle$, where m is the product of the different primes dividing n. The Frattini subgroups of the infinite cyclic group, S_n and $Z(p^\infty)$ are 1, 1, and $Z(p^\infty)$, respectively. Using the notation of Problem 1.6, we have $\Phi(D_k) = \langle b^m \rangle$, where m is the product of the different primes dividing k. Using the notation of Problem 5.29, we have $\Phi(G) = \langle u^2 \rangle$.

8.2. Let us suppose that $x_1, \ldots, x_n$ generate G. Since $G = H\Phi(G)$, we have $x_i = y_i u_i$ $(y_i \in H, u_i \in \Phi(G))$ for each i. Therefore, using 8.T.1,

$$G = \langle y_1, \ldots, y_n, u_1, \ldots, u_n \rangle = \langle y_1, \ldots, y_n, u_1, \ldots, u_{n-1} \rangle$$
$$= \cdots = \langle y_1, \ldots, y_n \rangle \subseteq H.$$

Hence $G = H$.

8.3. Since G is nilpotent, $G' \subseteq \Phi(G)$, by 8.T.2. If $G' = \langle x_1, x_2, \ldots, x_k \rangle$, then, using 8.T.1, we have

$$HG' = \langle H, x_1, x_2, \ldots, x_k \rangle = \langle H, x_2, \ldots, x_k \rangle = \cdots = \langle H, x_k \rangle = H.$$

8.4. Let M be any maximal subgroup of G. Since $MG' \neq G$, we must have $G' \subseteq M$. Thus G' is contained in the intersection $\Phi(G)$ of all maximal subgroups of G. Hence G is nilpotent by 8.T.3.

8.5. We suppose that, for each $x \in G$, there exists $u \in \Phi(G)$ such that $x^{-1}Hx = u^{-1}Hu$; that is, $x \in N(H;G)u \subseteq N(H;G)\Phi(G)$. Then $G = N(H;G)\Phi(G)$. By Problem 1.36 and 8.T.4, we conclude that $G = N(H;G)$, and so H is normal in G.

8.6. Let P be any Sylow p-group of $\Phi(G)$. For each $x \in G$, we have $x^{-1}Px \subseteq x^{-1}\,\Phi(G)x = \Phi(G)$, and so $x^{-1}Px$ is also a Sylow p-group of $\Phi(G)$. By Problem 2.23, this implies that P is conjugate to $x^{-1}Px$ in $\Phi(G)$, and so P is normal in $\Phi(G)$, by Problem 8.5. Thus all the Sylow subgroups of $\Phi(G)$ are normal in $\Phi(G)$, and so $\Phi(G)$ is nilpotent, by Problem 6.6. (For related results see [11]. Compare with Problem 8.32.)

8.7. Let $x_1\Phi, x_2\Phi, \ldots, x_k\Phi$ be a set of generators for G/Φ. Writing $H = \langle x_1, x_2, \ldots, x_k \rangle$, we have $G = H\Phi$. Hence, by 8.T.4, the elements $x_1, x_2, \ldots, x_k$ generate G. Conversely, if $y_1, y_2, \ldots, y_r$ generate G, then certainly $y_1\Phi, y_2\Phi, \ldots, y_r\Phi$ generate G/Φ. Hence G has no set of generators consisting of fewer than k elements.

8.8. Let us suppose that P was contained in $\Phi(G)$. Then P is the unique Sylow p-group of $\Phi(G)$, by Problems 8.6 and 6.10, and so P is a characteristic subgroup of $\Phi(G)$. Therefore, by Problem 3.20, P is normal in G. Hence, by Problem 7.6, P has a complement K in G. Let M be a maximal subgroup of G such that M contains K. Clearly, P does not lie in M, and so P cannot lie in $\Phi(G)$, which is contained in every maximal subgroup of G. This gives the required contradiction.

8.9. First, let us consider the case where H is a normal subgroup of G. Then $\Phi(H)$ is a normal subgroup of G, by Problem 3.20. Now we suppose that the assertion is false; that is, $\Phi(H)$ does not lie in $\Phi(G)$. Then there exists a maximal subgroup M of G such that M does not contain $\Phi(H)$. Hence $G = M\Phi(H)$, and so $H = H \cap M\Phi(H) = (H \cap M)\Phi(H)$, by Problem 1.38. By 8.T.4 and Problem 1.36, this implies that $H \cap M = H$, and so $H \subseteq M$. This contradicts our hypothesis, and so $\Phi(H) \subseteq \Phi(G)$ as asserted.

More generally, if H is a subnormal subgroup of G, let

$$G = H_0 \supseteq H_1 \supseteq \cdots \supseteq H_k = H$$

be a normal series from G to H. Then, from the result above, we have

$$\Phi(G) = \Phi(H_0) \supseteq \Phi(H_1) \supseteq \cdots \supseteq \Phi(H_k) = \Phi(H).$$

8.10. If M is a maximal subgroup of G, then M/N is a maximal subgroup of G/N. [$N \subseteq M$ because $N \subseteq \Phi(G)$.] Conversely, every maximal subgroup L/N of G/N corresponds to a maximal subgroup L of G. The result now follows from the definition of the Frattini subgroup (see [76] and [77]).

8.11. Take $N = \langle x^m \rangle$, where the square of some prime divides m (see Problem 8.1).

8.12. Let G be the group defined in Problem 6.47. Then $\Phi(G) = A$, and so $| G:\Phi(G) | = 2$, but G is not finitely generated.

8.13. Let G be the symmetric group S_4, and take $H = \langle (1234) \rangle$. Then $\Phi(G) = 1$, but $\Phi(H) = \langle (13)(24) \rangle$.

8.14. If M/C is a maximal subgroup of G/C, then M is a maximal subgroup of G, and so $F \subseteq M$. Thus $CF/C \subseteq \Phi(G/C)$. On the other hand, if F were isomorphic to the quaternion group, then $CF/C \simeq F/(F \cap C) = F/Z(F)$ would be a noncyclic group of order 4. But G/C must be isomorphic to a subgroup of the symmetric group S_4, by Problems 3.17 and *3.40, and it is readily verified that every subgroup of S_4 has a Frattini subgroup of order 1 or 2. This gives the required contradiction. (For more general examples, see [5], page 163.)

8.15. Let

$$G = M_0 \supset M_1 \supset \cdots \supset M_k = M$$

be a minimal normal series from G to M. We suppose that M is not normal in G. Then $k \geqslant 2$ and, for some $x \in M_{k-2}$, we have $x^{-1}Mx \neq M$. But $x^{-1}Mx$ is a normal subgroup of M_{k-1}, and so $Mx^{-1}Mx$ is a subnormal subgroup of G properly containing M. Furthermore, it follows from Problem 3.20 that $Mx^{-1}Mx$ is p-closed. This contradicts the choice of M.

8.16. Put $\Phi = \Phi(G)$, and let P be a Sylow p-group of M. Since M/Φ is p-closed and normal in G/Φ, Problem 3.20 shows that $P\Phi/\Phi$ is a normal subgroup of G/Φ, and so $x^{-1}Px \subseteq P\Phi$ for all $x \in G$. Because P and $x^{-1}Px$ are both Sylow p-groups of $P\Phi$, Problem 2.23 shows that, for each $x \in G$, there exists $u \in \Phi$ such that $x^{-1}Px = u^{-1}Pu$. Then Problem 8.5 shows that P is normal in G, and, in particular, M is p-closed.

8.17. By hypothesis, the group $H\Phi(G)/\Phi(G)$, which is isomorphic to $H/(H \cap \Phi(G))$, is a p-closed subnormal subgroup of $G/\Phi(G)$ such that

M contains $H\Phi(G)$. Using Problems 8.15 and 8.16, we have that M is a normal subgroup of G and so is p-closed. Let P be its unique Sylow p-group. Then $P \cap H$ is the unique Sylow p-group of H.

8.18. If H is nilpotent, then by 8.T.2 and Problem 8.9 we have $H' \subseteq \Phi(H) \subseteq \Phi(G)$.

Conversely, if $H' \subseteq \Phi(G)$, then $H' \subseteq H \cap \Phi(G)$. Hence $H/(H \cap \Phi(G))$ is abelian by 5.T.1, and so is p-closed for each prime p. Thus, by Problem 8.17, H has a unique Sylow p-group for each prime p. By Problem 6.6 this implies that H is nilpotent. (For the results of Problems 8.15 to 8.18 see [78]. The result of Problem 8.18 is due to Gaschütz [76].)

8.19. Let us suppose that $N \subseteq \Phi(U) \subseteq U$ but N does not lie in $\Phi(G)$. Then we can choose a maximal subgroup M of G such that N does not lie in M. Then $G = MN$, and so $U = U \cap MN = (U \cap M)N$, by Problem 1.38. Therefore $U = (U \cap M)\Phi(U)$, and so $U = U \cap M$, by Problem 8.2. Thus $N \subseteq U \subseteq M$, contrary to the choice of M.

8.20. From the definition of $\Phi(G)$ it is sufficient to prove that each maximal subgroup M of G contains $Z(G) \cap G'$. Since M is maximal, $Z(G) \subseteq M$ or $G = MZ(G)$. In the latter case, M is normal and G/M is abelian, so $M \supseteq G'$. Hence, in either case, $Z(G) \cap G' \subseteq M$.

8.21. If M is a maximal subgroup of A, then $M \times B$ is a maximal subgroup of G. Therefore $\Phi(G) \subseteq \Phi(A) \times B$. Using a similar inequality for B, we have $\Phi(G) \subseteq (\Phi(A) \times B) \cap (A \times \Phi(B)) = \Phi(A) \times \Phi(B)$.

8.22. Let $a_1b_1, \ldots, a_nb_n$ be a set of generators for G with $a_i \in A$ and $b_i \in B$ for each i. Then $a_1, \ldots, a_n$ and $b_1, \ldots, b_n$ generate A and B, respectively. In particular, A and B are finitely generated.

We now show that $\Phi(A)$ and $\Phi(B)$ are both contained in $\Phi(G)$. Otherwise, there exists a maximal subgroup M of G such that $\Phi(A)$, say, is not contained in M. Since $\Phi(A)$ is normal in G, by Problem 3.20, $G = M\Phi(A)$, and so $A = A \cap M\Phi(A) = (A \cap M)\Phi(A)$, by Problem 1.38. Since A is finitely generated, this implies $A = A \cap M$, by Problem 8.2, and hence $A \subseteq M$, contrary to the choice of M. Thus we conclude that $\Phi(A) \times \Phi(B) \subseteq \Phi(G)$, and then the result follows from Problem 8.20. (Both Problems 8.20 and 8.21 may be generalized to direct products of infinitely many subgroups. Moreover, the condition in Problem 8.21 that G be finitely generated may be relaxed to some extent. For further details see [79].)

8.23. Now $H \cap A$ is normal in H, because A is normal in G, and it is normal in A, because A is abelian. Therefore $H \cap A$ is normal in $G = HA$.

8.24. We proceed by induction on $|A|$. Since $A \cap \Phi(G) = 1$, there is a maximal subgroup M of G not containing A, and then $G = MA$. Furthermore, $B = M \cap A$ is a proper subgroup of A which is normal in G, by Problem 8.23. If $B = 1$, we take $K = M$. Otherwise, by the induction hypothesis, there is a subgroup U of G such that $G = UB$ and $U \cap B = 1$. Then $A = A \cap UB = (A \cap U)B$, by Problem 1.38. Since $A \cap U$ is a proper subgroup of A, and $A \cap U$ is normal in G by Problem 8.23, we can apply the induction hypothesis to find a subgroup V of G such that $G = V(A \cap U)$ and $A \cap U \cap V = 1$. Finally, $U = U \cap V(A \cap U) = (U \cap V)(A \cap U)$, by Problem 1.38. Therefore, with $K = U \cap V$, we find $G = UA = K(A \cap U)A = KA$ and $A \cap K = 1$ (see [76]).

8.25. We have that Φ is a characteristic subgroup of G. Thus, for each $\alpha \in \text{Aut } G$, there is an induced automorphism $\bar{\alpha} \in \text{Aut } (G/\Phi)$. It is easy to see that the mapping $\alpha \to \bar{\alpha}$ defines a homomorphism of Aut G onto Aut (G/Φ). Therefore, in order to prove the required result, we must show that the kernel K of this homomorphism has order dividing $|\Phi|^k$.

Let $u_1, u_2, \ldots, u_k$ be a set of generators for G. Then an automorphism of G lies in K if and only if $\Phi u_i^\alpha = \Phi u_i$; that is, $u_i^\alpha u_i^{-1} \in \Phi$, for $i = 1, 2, \ldots, k$.

For any elements a_i in Φ, the elements

$$(*) \qquad a_1u_1, a_2u_2, \ldots, a_ku_k$$

also generate G. In fact, using 8.T.1, we have

$$\begin{aligned} G &= \langle u_i \mid i = 1, 2, \ldots, k \rangle = \langle a_iu_i, a_i \mid i = 1, 2, \ldots, k \rangle \\ &= \langle a_iu_i \mid i = 1, 2, \ldots, k \rangle. \end{aligned}$$

Corresponding to different choices of the a_i, we have a set Ω consisting of $|\Phi|^k$ different sets of generators of G of the form (*). Because $u_i^\alpha u_i^{-1} \in \Phi$ for all $\alpha \in K$, the set Ω is permuted by the mappings

$$\alpha^*: (a_1u_1, a_2u_2, \ldots, a_ku_k) \to (b_1u_1, b_2u_2, \ldots, b_ku_k) \qquad (\alpha \in K),$$

where $b_i = a_i^\alpha u_i^\alpha u_i^{-1} \in \Phi$. Thus K acts as a permutation group on Ω. Since no set of generators for G can be left unchanged by a nontrivial automorphism of G, the orbits under K each have length $|K|$ (see Problem 2.38). Therefore, by Problem 2.9, $|K|$ divides $|\Omega| = |\Phi|^k$. Hence $|\text{Aut } G| = |\text{Aut } (G/\Phi)|\,|K|$ divides $|\Phi|^k\,|\text{Aut } (G/\Phi)|$ (see [9], page 52).

8.26. Let us write $\Phi = \Phi(G)$. By 8.T.2, $G' \subseteq \Phi$, and so G/Φ is abelian. Since G is a p-group, each maximal subgroup M is a normal subgroup of index p. Therefore, for each $x\Phi \in G/\Phi$, we have $(x\Phi)^p = x^p\Phi \subseteq M$. Since Φ is the intersection of all maximal subgroups of G, $(x\Phi)^p \subseteq \Phi$. Thus G/Φ is an elementary abelian p-group.

8.27. Since an elementary abelian p-group of order p^k has a set of k generators, it follows from Problems 8.26 and 8.7 that G can be generated by a set of k generators. Therefore, by Problem 8.25, $|\operatorname{Aut} G|$ divides $|\Phi|^k |\operatorname{Aut}(G/\Phi)| = p^{k(n-k)} |\operatorname{Aut}(G/\Phi)|$. The result now follows from Problems 3.35 and 8.26 (see [59]).

8.28. Using Problem 6.36, we choose the subgroup A with the property that $C(A;G) = A$. By Problem 3.17, $|G:A|$ divides $|\operatorname{Aut} A|$, and so, from Problem 8.27, we conclude that p^{n-m} divides $p^{m(m-1)/2}$. Hence $n - m \leqslant \frac{1}{2}m(m-1)$; that is, $m(m+1) \geqslant 2n$. (See [9], page 145, and see [80] for related results.)

8.29. It is sufficient to show that each maximal subgroup M of prime index in G contains G''. We shall proceed by induction on $|G|$, and note that the result is trivial if $|G|$ is prime. Consider two cases:

(a) If M contains a nontrivial normal subgroup N of G, then, using 5.T.3, $(G/N)'' = G''N/N \subseteq M/N$ by the induction hypothesis. Hence $G'' \subseteq M$.

(b) If M contains no nontrivial normal subgroup of G, let A be a minimal normal subgroup of G. Then A is abelian, by Problem 6.11. Since A does not lie in M, therefore $G = AM$. Let $C = C(A;G)$ which is normal in G, by Problem 1.14. If $K = M \cap C$, then the normal closure K^G of K in G equals $K^{AM} = K^M \subseteq M$. Since M contains no nontrivial normal subgroup of G, this means that $K^G = K = 1$. Therefore $G/C = MC/C \simeq M/(M \cap C) = M$; thus, $|A|$ is prime, and so, by Problem 3.18, M is an abelian group. This implies that G/A, which is isomorphic to M, is also abelian. Therefore $G' \subseteq A$, and so $G'' = 1$. Thus $G'' \subseteq M$ as required.

8.30. We proceed by induction on $|G|$. If $N \subseteq \Phi(G)$, then we may take $H = G$ because N is nilpotent, by Problem 8.5. If $\Phi(G)$ does not contain N, then some maximal subgroup M of G does not contain N. By induction, there exists a subgroup H of M such that $M = H(N \cap M)$ and $H \cap (N \cap M) = H \cap N$ is nilpotent. Then $G = HN$ as required.

8.31. Let $x_1, x_2, \ldots, x_n$ be a fixed set of coset representatives for N in G. For each $\alpha \in A$ there exist $u_i \in N$ $(i = 1, 2, \ldots, n)$ such that $x_i^\alpha = u_i x_i$. For all $y \in N$, we have $x_i^{-1} y x_i = (x_i^{-1} y x_i)^\alpha = (x_i^{-1})^\alpha y^\alpha x_i^\alpha = x_i^{-1} u_i^{-1} y u_i x_i$.

Thus $y = u_i^{-1} y u_i$ for each $y \in N$, and so each of the elements u_i lies in $Z(N)$. Finally, we note that the mapping of A into B defined by $\alpha \rightarrow (u_1, u_2, \ldots, u_n)$ $(\alpha \in A)$ is a homomorphism, and the kernel of the homomorphism is 1. Thus A is isomorphic to a subgroup of B. (For more general results see [60].)

8.33. First suppose that A is a finite cyclic group of order m, say. If $p_1, \ldots, p_k$ are the different primes which divide m, then it follows from Problem 8.1 that a cyclic group of order $mp_1 \ldots p_k$ has its Frattini subgroup isomorphic to A.

In the general case we can write $A = C_1 \times \cdots \times C_s$ with each C_i a cyclic subgroup of A, using 3.T.5. Then, for each i, we can use the result above to find a cyclic group B_i with $\Phi(B_i) \simeq C_i$. Finally, taking B as the external direct product $B_1 \times \cdots \times B_s$, we have $\Phi(B) \simeq A$ by Problem 8.22.

Solutions for Chapter 9—Factorization

9.1. Take $H = \langle(12345), (2354)\rangle$.

9.2. Clearly, a subgroup H of A_n is a complement of A_{n-1} if and only if H is a transitive subgroup of order n; that is, H is a regular subgroup of A_n. We consider three cases.

(a) If n is odd, then $(12 \ldots n)$ is an even permutation, and so $\langle(12 \ldots n)\rangle$ is the required complement.

(b) If $n = 2m$ (m odd), then any regular permutation group on n letters contains an element of order 2 of the form $(\xi_1\xi_2)(\xi_3\xi_4) \cdots (\xi_{n-1}\xi_n)$, which is an odd permutation (compare with Problem 2.16). Thus, in this case there is no complement.

(c) If $n = 2m$ (m even), then

$$\langle(1\ 2 \ldots m)(m+1\ m+2 \ldots 2m),\ (1\ m+1)(2\ m+2) \cdots (m\ 2m)\rangle$$

is a complement of A_{n-1} in A_n.

9.3. Let G be the group generated by x, y, z, and w subject to the relations

$$x^3 = y^3 = z^2 = w^2 = 1, \qquad [x,y] = [z,w] = [z,x] = [w,y] = 1,$$

and

$$z^{-1}yz = y^{-1}, \qquad w^{-1}xw = x^{-1}$$

Then $A = \langle x,z \rangle$ and $B = \langle y,w \rangle$ are cyclic groups of order 6. Moreover, $G = AB$ but neither A nor B is normal in G.

9.4. Take $\langle a \rangle$ and $\langle ab \rangle$ in the dihedral group D_4 defined in Problem 1.6.

9.5. If $AB = BA$, then for $a,a_1 \in A$ and $b,b_1 \in B$ we have

$$(ab)(a_1b_1) = a(ba_1)b_1 = aa_2b_2b_1 \in AB$$

and

$$(ab)^{-1} = b^{-1}a^{-1} = a_3b_3 \in AB$$

for some a_2, $a_3 \in A$ and b_2, $b_3 \in B$. Thus AB is a subgroup and clearly equals $\langle A,B \rangle$.

Conversely, if $AB = \langle A,B \rangle$, then $AB = \{(ab)^{-1} \mid a \in A \text{ and } b \in B\} = \{b^{-1}a^{-1} \mid a \in A \text{ and } b \in B\} = BA$.

9.6. Let $z = ab$ be an element of $Z(G)$ with $a \in A$ and $b \in B$. Then $a = zb^{-1}$ commutes with all elements of A, because A is abelian, and it commutes with all elements of B, because B is abelian and $z \in Z(G)$. Thus $a \in Z(G)$. Similarly, $b \in Z(G)$. Therefore $Z(G)$ is contained in $(A \cap Z(G))(B \cap Z(G))$. The reverse inequality is trivial, so the result follows.

9.7. Write $x = ab$ with $a \in A$ and $b \in B$. Then $(x^{-1}Ax)B = b^{-1}AbB = b^{-1}ABb = G = AB$. Similarly, $(x^{-1}Ax)(y^{-1}By) = (x^{-1}Ax)B = G$.

9.8. If $B = x^{-1}Ax$, then by Problem 9.7 we have $G = Ax^{-1}Ax = AA = A$. Thus A is not a proper subgroup of G.

9.9. If G is nilpotent, then every maximal subgroup of G is normal, by Problem 6.9, and so every pair of maximal subgroups commute. Conversely, let us suppose that every pair of maximal subgroups of G commute. If M is a maximal subgroup, then $Mx^{-1}Mx \neq G$ for any $x \in G$, by Problem 9.8. Thus $M = Mx^{-1}Mx$; that is, $x^{-1}Mx \subseteq M$ for all $x \in G$. Hence M is normal in G (see [19]).

9.10. We proceed by induction on the index $|G{:}A|$. If A is normal in G, then there is nothing to prove; so we suppose that some $x \in G$

does not normalize A. Then A is a proper subgroup of the group $B = Ax^{-1}Ax$, and, by Problem 9.8, B is a proper subgroup of G. Because A commutes with each of its conjugates in G, the same is true for B. Thus, from the induction hypothesis, A is subnormal in B, and B is subnormal in G. Therefore A is subnormal in G. (See [19]. For related results see [81].)

9.11. Using the generic notation $a_i \in A$ and $b_i \in B$, we have

$$a_2(a_j^{-1}b_1^{-1}a_jb_1)a_2^{-1} = a_j^{-1}b_3^{-1}a_jb_3 \qquad \text{where } a_3b_3 = b_1a_2^{-1},$$

and

$$b_2(a_1^{-1}b_j^{-1}a_1b_j)b_2^{-1} = a_4^{-1}b_j^{-1}a_4b_j \qquad \text{where } b_4a_4 = a_1b_2^{-1}.$$

Therefore $b_2a_2[a_1, b_1]a_2^{-1}b_2^{-1} = a_4^{-1}b_3^{-1}a_4b_3 = a_2b_2[a_1, b_1]b_2^{-1}a_2^{-1}$. Thus $[a_1, b_1]$ commutes with $[a_2, b_2]$.

Let K be the set union of A and B. Then $[K,K] = [A,B]$ because A and B are abelian. Since $[A,B]$ is a normal subgroup of G, by Problem 5.4, therefore $G' = [A,B]$, by Problem 5.2. Since we have shown that $[A,B]$ is abelian, it follows that $G'' = 1$ (see [82]).

9.12. Let $N \subseteq A \cap B$ be a normal subgroup of B. Then the normal closure $N^G = N^{BA} = N^A \subseteq (A \cap B)^A \subseteq A$; so N^G is the required normal subgroup.

9.13. If, for some $x \in G$, we have $B \cap x^{-1}Bx = N \neq 1$, then N is a nontrivial normal subgroup of B. Since $G = x^{-1}AxB$, by Problem 9.7, therefore A contains a nontrivial normal subgroup of G, by Problem 9.12.

9.14. We proceed by induction on $|G|$. Consider two cases:

(a) *There exists a nontrivial normal subgroup M of G such that $AM \neq BM$.* Then, by the induction hypothesis applied to $G/M = (AM/M)(BM/M)$, we can find a proper normal subgroup N/M of G/M containing AM/M, say. Then $N \supseteq A$.

(b) *For all nontrivial normal subgroups M of G, $AM = BM$, and so both these subgroups equal G.* Because of Problem 9.13, the hypothesis implies that $A \cap B = 1$. We note that G is solvable by Problem 9.11, and we take M to be a minimal normal subgroup of G. Then, using 1.T.2, we have $|G| = |G:A|\,|G:B| = |AM:A|\,|BM:B| = |M:A \cap M|\,|M:B \cap M|$. Since M is a p-group by Problem 6.11, G is a p-group. Let us suppose that $A \neq G$. Then any maximal subgroup N of G which contains A is normal in G because G is nilpotent (see Problem 6.8). This proves the result in this second case.

9.15. We proceed by induction on $|G|$. By Problem 9.14 we may suppose that A is contained in a proper normal subgroup of G, and so the normal closure $K = A^G$ of A is a proper subgroup of G. From Problem 5.13 we find that $K = A[A,G] = A[A,B] = AG'$. Since $K = K \cap AB = A(K \cap B)$ by Problem 1.38, the induction hypothesis implies that there exists at least one nontrivial normal subgroup of K contained in either A or $B \cap K$. Consider two cases.

(a) *Some nontrivial normal subgroup N of K lies in $K \cap B$.* Then N is normalized by both A and B. Hence N is a normal subgroup of $G=AB$ and N is contained in B.

(b) *B contains no nontrivial normal subgroup of K, and so some normal subgroup $N \neq 1$ of K is contained in A.* We first show that this implies that $Z(K) \neq 1$. In fact, if $A \cap G' \neq 1$, then $A \cap G'$ is a nontrivial subgroup of $Z(K)$ because A and G' are abelian, by Problem 9.11. On the other hand, if $A \cap G' = 1$, then $N \cap K' \subseteq A \cap G' = 1$, and so $N \subseteq Z(K)$, by Problem 5.12. Thus, in either case, $Z(K) \neq 1$. Since $K = A(B \cap K)$, Problem 9.6 implies that $Z(K) = (A \cap Z(K))(B \cap Z(K))$. By hypothesis, the normal subgroup $B \cap Z(K)$ of K in B is trivial, and so $Z(K) \subseteq A$. Finally, since $Z(K)$ is a characteristic subgroup of K, it is normal in G, by Problem 3.20. (For the results of Problems 9.14 and 9.15 see [82].)

9.16. Let $ab \in H$ with $a \in A$ and $b \in B$. Choose integers k and l such that $km + ln = 1$. Then $a = a^{1-km} = a^{ln} \in Hb^{-ln} = H$, and similarly, $b \in H$. It follows that $H \subseteq (A \cap H)(B \cap H)$, and so $H = (A \cap H)(B \cap H)$ as asserted.

9.17. We have $|A:A \cap B| = |G:B|$, by 1.T.2. Since N is normal in A, and the order of N is relatively prime to the index of $A \cap B$ in A, it follows from Problem 1.18 that $N \subseteq A \cap B$. Thus $N^G = N^{AB} = N^B \subseteq B$.

9.18. We proceed by induction on $|G|$. By Problem 6.11, A has a minimal normal subgroup N whose order is a prime power, say p^n. Since $|G:B|$ and $|G:C|$ are relatively prime, at least one of these indices is prime to p. We suppose that p does not divide $|G:B|$. By Problem 1.8, $G = AB$. Thus, by Problem 9.17, the normal subgroup $M = N^G$ of G lies in B. Since B is solvable, M is solvable. On the other hand, applying the induction hypothesis to G/M with subgroups AM/M, B/M, and CM/M, whose indices are relatively prime in pairs, we deduce that G/M is solvable. Hence G is solvable. (See [83]. A related result appears in [84].)

9.19. Clearly G is an M-group by Problem 1.38. To prove the second part we shall show that, for any proper subgroup H of G, and any $x \notin H$, H is properly contained in its normalizer in $H\langle x\rangle$. Consider two cases:

(a) *For some integer* $m > 1$, $H \cap \langle x\rangle = \langle x^m\rangle$. Then $|H\langle x\rangle : H| = |\langle x\rangle : \langle x^m\rangle|$, which is finite. Therefore H is subnormal in $H\langle x\rangle$, by Problem 9.10, and so it cannot be its own normalizer.

(b) $H \cap \langle x\rangle = 1$ *and* x *has infinite order.* Then each element of $H\langle x\rangle$ has a unique representation in the form ux^r ($u \in H$, r integer). In fact, $ux^r = vx^s$ ($v \in H$) implies that $u^{-1}v = x^{r-s} \in H \cap \langle x\rangle = 1$. Furthermore, for any prime p, $H\langle x^p\rangle$ is a maximal subgroup of index p in $H\langle x\rangle$. Since $H\langle x^p\rangle$ is subnormal in $H\langle x\rangle$, by Problem 9.10, therefore it is normal and so contains the derived group of $H\langle x\rangle$, by 5.T.1. Thus, for each $u \in H$, there exists $v \in H$ and an integer r such that $u^{-1}x^{-1}ux = vx^{pr}$. This is true for each prime p. But the representation of elements in this form is unique, so $pr = 0$. Hence $[u,x] = v \in H$; that is, $x^{-1}ux \in H$, for all $u \in H$. Therefore H is normal in $H\langle x\rangle$.

9.20. Let us suppose that every pair of cyclic subgroups of G commute. Then, for any pair of subgroups S and T of G, we have, for all $s \in S$ and $t \in T$, $ts \in \langle t\rangle\langle s\rangle = \langle s\rangle\langle t\rangle \subseteq ST$. Thus $TS \subseteq ST$. Similarly, $ST \subseteq TS$, and so $ST = TS$.

9.21. Because of Problem 9.20, it is sufficient to show that any two cyclic subgroups A and B of G commute. We define a sequence of subgroups of B by:

$$B_0 = A \cap B \qquad \text{and} \qquad B_n = N(AB_{n-1};B) \qquad (n = 1, 2, \ldots).$$

Then $B_{n-1} \subseteq B_n$, and so $AB_n = (AB_{n-1})B_n = B_n(AB_{n-1})$. Thus it follows, by induction on n, that A and B_n commute for all $n \geqslant 0$. From the hypothesis (*), $B_n = N(AB_{n-1};B) \neq AB_{n-1} \cap B = B_{n-1}$ unless $B_{n-1} = B$. Hence we have a proper ascending chain, $B_0 \subset B_1 \subset B_2 \subset \cdots$, which terminates only if B is reached. But B is cyclic, so the chain is finite, by Problem 1.36. Hence $B_k = B$ for some $k \geqslant 0$, and so B commutes with A.

9.22. Let A be any proper subgroup of H. We must show that $N(A;H) \neq A$. Let $\mathcal{S}$ be the (nonempty) set of subgroups B of G such that $B \cap H = A$. If $\mathfrak{I}$ is any totally ordered subset of $\mathcal{S}$ (that is, for any B and C in $\mathfrak{I}$, either $B \subseteq C$ or $C \subseteq B$), then $\bigcup_{B\in\mathfrak{I}} B$ is a subgroup lying in $\mathcal{S}$ and containing each of the subgroups in $\mathfrak{I}$. Thus we may apply Zorn's lemma to obtain a maximal element M of $\mathcal{S}$. By hypothesis, $N = N(M;G)$ properly contains M (since $M \neq G$), and so $N \notin \mathcal{S}$. But $A =$

$M \cap H$ is a normal subgroup of $N \cap H$. Therefore $N(A;H) \supseteq N \cap H \supset A$. (Compare with the proof in [2], Section 63.)

9.23. Let A and B be any subgroups of G such that $N(A;B) = A \cap B$. Since G is an M-group, $\langle A,B\rangle \cap N(A;G) = \langle A,N(A;G) \cap B\rangle = A$. But $\langle A,B\rangle$ satisfies the normalizer condition, by Problem 9.22, and so $A = \langle A,B\rangle$. Thus $B \subseteq A$. The result now follows from Problem 9.21. (For related results see [85].)

9.24. Take G as the symmetric group S_4, and let $A = S_3$, $B = \langle(1234)\rangle$, and $N = A_4$.

***9.25.** A finite group G has the desired property if and only if it is a direct product of subgroups each of which has squarefree order, that is, order of the form $p_1p_2 \cdots p_k$, where the p_i are distinct primes (see [86]).

Solutions for

Chapter 10—Linear Groups

10.1. Let $\gamma \in \mathcal{C}$ be an eigenvalue for c, and put $\mathcal{W} = \mathcal{V}(c - \gamma 1)$. Then, for each $x \in G$, $\mathcal{W}x = \mathcal{W}$ because $x(c - \gamma 1) = (c - \gamma 1)x$. Thus $\mathcal{W}$ is an invariant subspace for G, and, since $c - \gamma 1$ is not invertible, $\mathcal{W} \neq \mathcal{V}$. Since G is irreducible, $\mathcal{W} = 0$; that is, $c - \gamma 1 = 0$ as required. This proves the first part. The second part now follows because a group of scalars is irreducible only if it has degree 1.

10.2. Take $\mathbf{G} = TL(2)$.

10.3. If G were reducible, then $\mathcal{V} = \mathcal{W}_1 \dotplus \mathcal{W}_2$, where $\mathcal{W}_1$ and $\mathcal{W}_2$ are nontrivial invariant subspaces for G. Then $c \in GL(\mathcal{V})$ will commute with all the elements of G if $c|\mathcal{W}_1$ and $c|\mathcal{W}_2$ are scalars, and so there are nonscalars in $GL(\mathcal{V})$ which centralize G.

10.4. The first part is trivial. To prove the second part consider the groups $\mathbf{H}_m = \{[\xi_{ij}] \in STL(n) \mid \xi_{ij} = 0 \quad \text{whenever} \quad j < i \leqslant j + m\}$ $(m = 0, 1, \ldots)$. Then $\mathbf{H}_0 = STL(n)$ and $\mathbf{H}_{n-1} = \mathbf{1}$. Direct calculation now shows that $[\mathbf{H}_l, \mathbf{H}_m] = \mathbf{H}_{l+m+1}$ for all l and m, and so $^{n-1}(\mathbf{H}_0) = \mathbf{H}_{n-1}$. Hence $STL(n) = \mathbf{H}_0$ has class $n - 1$.

10.5. The set of all $n \times n$ matrices with entries in $\mathcal{C}$ is a vector space over $\mathcal{C}$ containing $\mathcal{C}\mathbf{S}$ as a subspace. Since the former space has dimension n^2 [it has a basis $\mathbf{e(i,j)}$ $(1 \leqslant i,j \leqslant n)$ where $\mathbf{e(i,j)}$ is the matrix with its (i,j)th entry 1 and the remaining entries 0], the dimension of $\mathcal{C}\mathbf{S}$ is at most n^2.

10.6. The mapping $x \to (x|\mathcal{W}_1, \ldots, x|\mathcal{W}_k)$ $(x \in G)$ is the required isomorphism (compare with Problem 2.36).

10.7. We first show that, for each $x \in G$, $\mathcal{W}x$ is a minimal invariant subspace for H. In fact, since H is normal, $\mathcal{W}xu = \mathcal{W}(xux^{-1})x = \mathcal{W}x$ for each $u \in H$, and so $\mathcal{W}x$ is invariant under H. Because $\mathcal{W}$ is a *minimal* invariant subspace for H, it follows that so is $\mathcal{W}x$.

We now choose successively elements $x_i \in G$ in the following manner. Take $x_1 = 1$, and for $i \geqslant 2$, choose x_i so that

$$\mathcal{W}x_i \cap (\mathcal{W}x_1 + \cdots + \mathcal{W}x_{i-1}) = 0.$$

Then $\mathcal{W}x_1 + \cdots + \mathcal{W}x_{i-1} + \mathcal{W}x_i$ is a direct sum. Since the dimensions of the subspaces are each at least 1, the process will terminate after, say, k steps.

Finally, $\mathcal{V} = \mathcal{W}x_1 \dot{+} \cdots \dot{+} \mathcal{W}x_k$. In fact, if $x \in G$, then by the choice of k, $\mathcal{W}x \cap (\mathcal{W}x_1 \dot{+} \cdots \dot{+} \mathcal{W}x_k) = \mathfrak{X}$, say, is nonzero. Since $\mathcal{W}x$ is a minimal invariant subspace for H, and $\mathfrak{X}$ is invariant for H, $\mathcal{W}x = \mathfrak{X}$, and so $\mathcal{W}x$ lies in the direct sum. Thus $\mathcal{W}x_1 \dot{+} \cdots \dot{+} \mathcal{W}x_k$ is invariant under G. Since G is irreducible, the result follows (compare with Problem 4.11).

10.8. With the notation of Problem 10.7, the mapping

$$u|\mathcal{W}x_i \to (x_j^{-1}x_iux_i^{-1}x_j)|\mathcal{W}x_j \qquad (u \in H)$$

is an isomorphism from $H|\mathcal{W}x_i$ onto $H|\mathcal{W}x_j$ for all i,j. Thus H has k isomorphic irreducible components. Since the dimensions of the corresponding subspaces are all equal, $k \mid n$. (See [87]. Compare with Problem 2.43.)

10.9. By Problem 10.8, A is completely reducible, so the result follows from Problem 10.1.

10.10. By Problem 10.1, there is a basis for $\mathcal{V}$ over which A corresponds to a diagonal group. Each of the basis elements is an eigenvector for each element of A and so lies in some root space. Hence $\mathcal{V} = \mathcal{R}_1 + \cdots + \mathcal{R}_k$. To show that this sum is direct, we must show that $(\mathcal{R}_1 + \cdots + \mathcal{R}_{i-1}) \cap$

$\mathcal{R}_i = 0$ for each i. We proceed by induction on i, noting that the result is true for $i = 1$. For $i \geqslant 2$, we have (by the induction hypothesis) that the sum $\mathcal{R}_1 \dot{+} \cdots \dot{+} \mathcal{R}_{i-1}$ is a direct sum. Now let us suppose that $v \neq \mathbf{0}$ lies in $(\mathcal{R}_1 \dot{+} \cdots \dot{+} \mathcal{R}_{i-1}) \cap \mathcal{R}_i$. Then $v = v_1 + \cdots + v_{i-1}$, where $v_j \in \mathcal{R}_j$ for $j = 1, 2, \ldots, i-1$. Choose j so that $v_j \neq 0$. Since $A|(\mathcal{R}_i + \mathcal{R}_j)$ is not a group of scalars (by the definition of a root space), therefore, for some $x \in A$, $x|\mathcal{R}_i = \alpha 1$ and $x|\mathcal{R}_j = \beta 1$, where the scalars α and β are different. Then

$$0 = \alpha v - vx = (\alpha v_1 - v_1 x) + \cdots + (\alpha v_j - v_j x) + \cdots + (\alpha v_{i-1} - v_{i-1} x).$$

Since we are dealing with a direct sum, this implies that $\alpha v_j - \beta v_j = 0$; that is, $v_j = 0$, contrary to the choice of j.

10.11. Since $A|\mathcal{R}_i$ is a group of scalars, and $x^{-1}Ax = A$ for all $x \in G$, therefore $A|\mathcal{R}_i x$ is also a group of scalars. Hence $\mathcal{R}_i x$ is contained in some root space $\mathcal{R}_j$. Similarly, $A|\mathcal{R}_j x^{-1}$ is a group of scalars. Since $\mathcal{R}_i \subseteq \mathcal{R}_j x^{-1}$, therefore $\mathcal{R}_i = \mathcal{R}_j x^{-1}$; that is, $\mathcal{R}_i x = \mathcal{R}_j$. Let

$$\Omega = \{\mathcal{R}_1, \mathcal{R}_2, \ldots, \mathcal{R}_k\}.$$

Then the mapping

$$x^*\colon \mathcal{R}_i \to \mathcal{R}_i x \qquad (i = 1, 2, \ldots, k)$$

is a permutation of Ω for each $x \in G$. The mapping $x \to x^*$ ($x \in G$) is a homomorphism of G into S_Ω, and the image G^* of this homomorphism is transitive because G is irreducible.

Finally, let us consider the kernel K of this homomorphism. If $z \in K$, then $A|\mathcal{R}_i$ is centralized by $z|\mathcal{R}_i$ for each i, and so $z \in C(A;G)$. Thus $K \subseteq C(A;G)$. Conversely, if $y \in C(A;G)$, then let $x \in A$ and suppose that the nonzero vector $v \in \mathcal{R}_i$ has an eigenvalue α for x. Then $(vy)x = vxy = \alpha(vy)$, and so vy is also an eigenvector with eigenvalue α for x. Thus $\mathcal{R}_i y = \mathcal{R}_i$ for each i, and so $y \in K$. Therefore $C(A;G) = K$. The result now follows. (For a more general result see [16], Sections 49 and 50.)

10.12. Let $x_1 = 1, x_2, \ldots, x_h$ be a set of right coset representatives for H in G. Let $\mathcal{W}$ be an invariant subspace for H for which $H|\mathcal{W}$ corresponds to a monomial group over a suitable basis. Clearly, $\mathcal{W}x_1 + \cdots + \mathcal{W}x_h$ is an invariant subspace for G, and since G is irreducible, this subspace is the whole of $\mathcal{V}$. Since the sum of the dimensions of the $\mathcal{W}x_i$ ($i = 1, \ldots, h$) equals the dimension of $\mathcal{V}$, the sum is a direct sum: $\mathcal{V} = \mathcal{W}x_1 \dot{+} \cdots \dot{+} \mathcal{W}x_h$. Thus, if $v_1, \ldots, v_m$ (with $m = n/h$) is a basis for $\mathcal{W}$ for which $H|\mathcal{W}$ corresponds to a monomial group, then $v_i x_j$ ($i = 1, \ldots, m$; $j = 1, \ldots, h$)

forms a basis for $\mathfrak{V}$. Finally, it is readily verified that G corresponds to a monomial group over this basis (see [88]).

10.13. We proceed by induction on the degree n of G. If G is reducible, then, since G is completely reducible by 10.T.1, each of the irreducible components of G has the given property, so the result follows from the induction hypothesis. If G is irreducible, and $n > 1$, then G is not abelian. Therefore, if we apply the given condition with $H = G$ and $N = 1$, we see that G has a normal abelian subgroup A not contained in $Z(G)$. Then, by Problem 10.11, $G/C(A;G)$ is isomorphic to a transitive permutation group of degree k, where $k \geqslant 2$ is the number of root spaces for A. With the notation of Problem 10.11, define $H = \{x \in G \mid \mathfrak{R}_1 x = \mathfrak{R}_1\}$. Then $H/C(A;G)$ is isomorphic to a stabilizer of G^* (see the proof of Problem 10.11), and so $| G{:}H | = k$, by Problem 2.30. Since all the root spaces for A have the same dimension, by Problem 10.11, therefore $H|\mathfrak{R}_1$ has degree n/k. Since $H|\mathfrak{R}_1$ has the given property of G, the induction hypothesis implies that $H|\mathfrak{R}_1$ corresponds to a monomial group over some basis for $\mathfrak{R}_1$. The result now follows from Problem 10.12.

10.14. (a) H/N is nilpotent. Let $Z/N = Z(H/N)$. If H/N is not abelian, then we can choose $x \in H$ such that xZ lies in the center of H/Z and $x \notin Z$. Then, using Problem 1.15, we find that $A = \langle Z,x \rangle$ is the required group.

(b) H/N is also solvable with its second derived group equal to 1. If H/N is not abelian, then $(H/N)' = NH'/N$, by 5.T.3, and is a nontrivial normal abelian subgroup of H/N. If $Z(H/N)$ contains NH'/N, then H/N is nilpotent of class 2, and the result follows from case (a). Otherwise, $A = H'N$ is the required subgroup.

(c) H/N is a finite solvable group which has all its Sylow subgroups abelian. Let $R/N = R(H/N)$ be the nilradical of H/N. By Problem 6.28, R/N is nilpotent. Since the Sylow subgroups of R/N are all abelian, it follows from Problem 6.10 that R/N is abelian. If H/N is not abelian, we shall show that R/N does not lie in the center of H/N. In fact, put $Z/N = Z(H/N)$. Then, since G/Z is solvable, there exists a nontrivial normal abelian subgroup U/Z of G/Z. Since $Z/N \subseteq Z(U/N)$, therefore U/N is nilpotent with class at most 2. Hence $U \subseteq R$, and R does not lie in Z. The result now follows with $A = R$. (See [16], Section 52; [89]; and [90].)

10.15. Because G is a nilpotent group, G corresponds to a monomial group over some basis $v_1, v_2, \ldots, v_n$, by Problem 10.14(a). Let Ω be the set of 1-dimensional subspaces $\mathfrak{V}_i$ spanned by v_i ($i = 1, 2, \ldots, n$).

Define $\mathcal{V}_i^x = \mathcal{V}_i x$ ($x \in G$; $\mathcal{V}_i, \mathcal{V}_i x \in \Omega$). Then G acts as a permutation group on Ω, and this group is transitive because G is irreducible. Thus $n = |\Omega|$ divides $|G|$ by Problem 2.12, and so n is a power of p.

10.16. By Problem 10.1, $Z(G)$ is a group of scalars, and hence is isomorphic to a subgroup Z of the multiplicative group of nonzero elements of $\mathcal{C}$. Since $Z(G)$ is a finite abelian group, Z may be written as a direct product of cyclic groups: $Z = C_1 \times C_2 \times \cdots \times C_s$, where the order of C_i divides the order of C_{i+1} (for $i = 1, 2, \ldots, s-1$), by Problem 3.4. Let us suppose that Z were not cyclic, and so $s > 1$. If $|C_1| = m$, then the equation $x^m = 1$ has $2m - 1$ distinct roots in the set $C_1 \cup C_2$. This means that the polynomial $x^m - 1$ has more than m roots in the field $\mathcal{C}$, which is not possible. Thus $s = 1$, and Z and $Z(G)$ are cyclic.

10.17. If G' is irreducible, then, by Problem 10.16, $Z(G')$ is cyclic. Hence G' is abelian by Problem 6.39. But, by Problem 10.1, an irreducible abelian group over an algebraically closed field has degree 1. Hence $n = 1$, contrary to hypothesis.

10.18. Since G is completely reducible, by 10.T.1, the solvable length of G is equal to the greatest solvable length of any of its irreducible components. Therefore it is sufficient to show that the result holds when G is irreducible and $l \geqslant 2$.

If G is irreducible, then $n = p^s$ for some integer $s > 0$, by Problem 10.15, and G' has solvable length $l - 1$. By Problem 10.17, G' is reducible, and its irreducible components will have degrees which are powers p^t of p, with $t < s$. Therefore, proceeding by induction on the solvable length of the group, we have $p^{l-2} \leqslant p^t \leqslant p^{s-1}$. The result then follows.

10.19. Let $x \in G$, and choose $\alpha 1 \in C^*$ such that $\alpha^n = \det x$. This is possible because $\mathcal{C}$ is algebraically closed. Then $x = (\alpha^{-1}x)(\alpha 1)$ is in G_1C^*. Thus $G \subseteq G_1C^*$. Since C^* is a group of scalars, G is irreducible if and only if GC^* is irreducible, and G_1 is irreducible if and only if G_1C^* is irreducible. Since $GC^* = G_1C^*$, the result follows.

10.20. Since G is irreducible, it follows from Problem 10.1 that $Z(G)$ is a group of scalars. But a scalar $\alpha 1$ lies in $SL(\mathcal{V})$ if and only if $\alpha^n = 1$. Thus $Z(G)$ is isomorphic to a subgroup of the group of nth roots of unity in $\mathcal{C}$.

10.21. We proceed by induction on k. The result holds for $k = 0$, by Problem 10.20. If $k > 0$, and $x \in Z_{k+1}$ and $y \in G$, then we have $[x,y] \in Z_k$, and so, by the induction hypothesis, $[x,y]^n \in Z_{k-1}$. The element xZ_{k-1} commutes with $[x,y]Z_{k-1}$ in G/Z_{k-1}. Therefore

$$x^{-n}y^{-1}x^ny Z_{k-1} = [x,y]^n Z_{k-1} = Z_{k-1}.$$

Thus, for all $y \in G$, $[x^n, y] \in Z_{k-1}$. Hence $x^n \in Z_k$ as asserted.

10.22. Let $\mathbf{x} = \text{diag}(\xi_1, \xi_2, \ldots, \xi_n)$. Then $\mathbf{x}^m = \mathbf{1}$ if and only if $\xi_1^m = \xi_2^m = \cdots = \xi_n^m = 1$. Thus there are m possible choices for each ξ_i in $\mathcal{C}$, and a total of m^n diagonal matrices $\mathbf{x}$ such that $\mathbf{x}^m = \mathbf{1}$. This set of matrices forms a group of which G is a subgroup. Therefore $|G|$ divides m^n.

10.23. It is sufficient to show that if P is a Sylow subgroup of G, then $|P|$ divides $n!m^n$. We proceed by induction on n. If P is reducible, then, by 10.T.1, it can be reduced into two components of degrees r and $n - r$, respectively, with $1 \leqslant r < n$. Then by Problem 10.6 and the induction hypothesis $|P|$ divides $r!m^r(n-r)!m^{n-r} = n!m^n \Big/ \binom{n}{r}$. Hence $|P|$ divides $n!m^n$. On the other hand, if P is irreducible, then P contains a normal self-centralizing abelian subgroup A, by Problem 6.36. Thus P/A is isomorphic to a transitive permutation group of degree $k \leqslant n$ by Problem 10.11, and so $|P{:}A|$ divides $n!$. But A corresponds to a diagonal group over a suitable basis, by Problem 10.9, and so $|A|$ divides m^n by Problem 10.22. Hence $|P|$ divides $n!m^n$ (see [33], Section 56).

10.24. By Problem 10.21, each element $x \in G$ satisfies $x^{n^k} = 1$. By Problem 6.36, G contains a self-centralizing normal abelian subgroup A, and so, as in the proof of Problem 10.23, $|G|$ divides $n!(n^k)^n = n!n^{kn}$. Finally, since $x^{n^k} = 1$ for all $x \in G$, the only primes p which divide $|G|$ must divide n. Since p^{n+1} does not divide $n!$ for any prime p, therefore $|G|$ divides $n^{kn}\prod_{p|n} p^n$, and so $|G|$ divides $n^{(k+1)n}$.

10.25. If we use the notation of Problem 10.19, G_1 is an irreducible nilpotent subgroup of class k in $SL(\mathcal{V})$, and so $|G_1|$ divides $n^{(k+1)n}$, by Problem 10.24. Since $|G{:}G \cap C^*| = |GC^*{:}C^*| = |G_1C^*{:}C^*| = |G_1{:}G_1 \cap C^*|$, and $G \cap C^* = Z(G)$, therefore $|G{:}Z(G)|$ divides $|G_1|$. The result now follows.

10.26. Let $\mathcal{V}$ be the underlying vector space of G. Then, because G is completely reducible, $\mathcal{V} = \mathcal{W}_1 \dotplus \mathcal{W}_2 \dotplus \cdots \dotplus \mathcal{W}_s$, where each $\mathcal{W}_i$ is a minimal invariant subspace for G. Let d_i be the dimension of $\mathcal{W}_i$ $(i = 1, 2, \ldots, s)$. If N_i is the subgroup of G consisting of all elements x

such that $x|\mathfrak{W}_i$ lies in the center of $G|\mathfrak{W}_i$, then, by Problem 10.25, $| G:N_i | \leqslant d_i^{(k+1)d_i} \leqslant n^{(k+1)d_i}$. Since

$$\bigcap_{i=1}^{s} N_i = Z(G),$$

therefore (see Problem 10.6)

$$| G:Z(G) | \leqslant \prod_{i=1}^{s} | G:N_i | \leqslant \prod_{i=1}^{s} n^{(k+1)d_i} \leqslant n^{(k+1)n}.$$

(See [91], Chapter III.)

10.27. First, let us suppose that G is irreducible. Using the notation of Problem 10.19, we have $G \subseteq G_1C^*$, where C^* is a normal abelian subgroup of G_1C^*. Hence it is sufficient to show that the solvable length m of G_1 satisfies $2^m \leqslant n$. Now G_1 is finite, by Problem 10.24, and so it is the direct product of its Sylow subgroups, by Problem 6.10. Hence, using Problem 10.18, we have that $p^{m_p} \leqslant n$, where m_p is the solvable length of the Sylow p-group of G. Since m is equal to the largest of the m_p (prime $p \mid n$), therefore $2^m \leqslant n$, and the result is proved in this case.

If G is reducible, then it can be reduced into two components of degrees r and $n - r$, say, with $1 \leqslant r < n$. At least one of these components is solvable of length l, and so the result follows by induction on the degree of the group.

10.28. If $\mathbf{y} \in \mathbf{H}$, then $\mathbf{y}$ has at most m conjugates $\mathbf{x}^{-1}\mathbf{y}\mathbf{x} = \mathbf{y}[\mathbf{x},\mathbf{y}]^{-1}$ $(\mathbf{x} \in \mathbf{G})$ in $\mathbf{G}$. Therefore $| \mathbf{G}:\mathbf{C}(\mathbf{y};\mathbf{G}) | \leqslant m$. From Problem 10.5, $\mathfrak{C}\mathbf{H}$ has dimension $s \leqslant n^2$, and so we can find elements $\mathbf{y}_1, \mathbf{y}_2, \ldots, \mathbf{y}_s$ in $\mathbf{H}$ which form a basis for $\mathfrak{C}\mathbf{H}$. Then

$$\mathbf{C}(\mathbf{H};\mathbf{G}) = \bigcap_{i=1}^{s} \mathbf{C}(\mathbf{y}_i;\mathbf{G}).$$

Hence

$$| \mathbf{G}:\mathbf{C}(\mathbf{H};\mathbf{G}) | \leqslant \prod_{i=1}^{s} | \mathbf{G}:\mathbf{C}(\mathbf{y}_i;\mathbf{G}) | \leqslant m^{n^2}.$$

(Compare with the proof of Problem 5.26.)

10.29. Let $Z = Z(G)$. Since G/Z is solvable, it contains a nontrivial normal abelian subgroup N/Z. Since $N' \subseteq Z \subseteq Z(N)$, therefore N is nilpotent of class at most 2. By Problem 10.8, N is completely reducible. The result now follows from Problems 10.26 and 10.20.

10.30. If G is reducible, there is nothing to prove. If G is irreducible, then, using the notation of Problem 10.19, we find that G_1 is an irreducible solvable subgroup of $SL(\mathcal{V})$. Let us choose N in G_1 as in Problem 10.29. Since $[G_1, N] \subseteq N$, and $|N| \leqslant n^{4n}$, therefore $|G_1:C(N;G_1)| \leqslant n^{4n^3}$, by Problem 10.28. Since N does not lie in $Z(G_1)$, N contains a nonscalar element, and so $C(N;G_1)$ is reducible, by Problem 10.1. Then $H = G \cap C(N:G_1)C^*$ is a reducible subgroup of G, and

$$|G:H| = |G_1C^*:C(N;G_1)C^*| \leqslant |G_1:C(N;G_1)| \leqslant n^{4n^3}.$$

Finally, since $C(N;G_1)$ is normal in G_1, therefore H is normal in G.

10.31. We proceed by induction on n. Let H be a reducible subgroup of G satisfying the condition of Problem 10.30. Then, since H is reducible, there is a basis for $\mathcal{V}$ such that for some r, $1 \leqslant r < n$, each element $x \in H$ corresponds to a matrix of the form

$$(*) \qquad \begin{bmatrix} \mathbf{x}_1 & \mathbf{0} \\ \mathbf{x}_3 & \mathbf{x}_2 \end{bmatrix}$$

where $\mathbf{x}_1, \mathbf{x}_3$, and $\mathbf{x}_2$ are blocks of dimensions $r \times r$, $(n-r) \times r$, and $(n-r) \times (n-r)$, respectively. The mappings $x \to \mathbf{x}_1$ $(x \in H)$ and $x \to \mathbf{x}_2$ $(x \in H)$, as defined by (*), are homomorphisms of H onto solvable groups $\mathbf{H}_1$ and $\mathbf{H}_2$, respectively, of lower degrees. Thus, by the induction hypothesis, there exist subgroups $\mathbf{M}_1$ and $\mathbf{M}_2$ of $\mathbf{H}_1$ and $\mathbf{H}_2$, respectively, which are similar to triangular groups, and such that $|\mathbf{H}_1:\mathbf{M}_1| \leqslant r^{4r^4}$ and $|\mathbf{H}_2:\mathbf{M}_2| \leqslant (n-r)^{4(n-r)^4}$. It then follows that the subgroup $M = \{x \in H \mid \mathbf{x}_1 \in \mathbf{M}_1$ and $\mathbf{x}_2 \in \mathbf{M}_2$ in $(*)\}$ of G corresponds to a triangular group over some basis of $\mathcal{V}$. Moreover,

$$|G:M| \leqslant |G:H|\,|\mathbf{H}_1:\mathbf{M}_1|\,|\mathbf{H}_2:\mathbf{M}_2| < n^{4n^3+4r^4+4(n-r)^4} < n^{4n^4}.$$

(See [12]; [68]; and [91], Theorem 15.)

10.32. From Problems 10.31 and 2.18, G contains a normal subgroup N whose index in G is at most $(n^{4n^4})!$, and N corresponds to a triangular group over some basis. From Problems 10.3 and 10.4 the solvable length of N is at most n, and so the solvable length of G is bounded above by $(n^{4n^4})! + n$. (See [92]. Huppert [14] shows that the solvable length l of a solvable linear group of degree n satisfies $2^{l-1} \leqslant n^7$.)

10.33. By Problem 6.36, G contains a self-centralizing normal abelian subgroup A, and G/A is isomorphic to a nilpotent permutation group of degree n, by Problem 10.11. From Problem 6.46, $|G:A| \leqslant 2^{n-1}$.

10.34. By induction on the degree of G. If G is irreducible, then the subgroup A defined in Problem 10.33 corresponds to a diagonal group over a suitable basis, by Problem 10.9, and so we may take $H = A$.

If G is reducible, then, as in the proof of Problem 10.31, we may suppose that each $x \in G$ corresponds to a matrix of the form (*) over a given basis. Let the images of the homomorphisms $x \to \mathbf{x}_1$ $(x \in G)$ and $x \to \mathbf{x}_2$ $(x \in G)$ be $\mathbf{G}_1$ and $\mathbf{G}_2$, respectively. Then, by the induction hypothesis, there is a subgroup $\mathbf{H}_i$ of $\mathbf{G}_i$ $(i = 1, 2)$ which is similar to a triangular group such that $| \mathbf{G}_1{:}\mathbf{H}_1 | \leqslant 2^{r-1}$ and $| \mathbf{G}_2{:}\mathbf{H}_2 | \leqslant 2^{n-r-1}$. Let $H = \{x \in G \mid \mathbf{x}_1 \in \mathbf{H}_1 \text{ and } \mathbf{x}_2 \in \mathbf{H}_2 \text{ in } (*)\}$. Then H corresponds to a triangular group over a suitable basis, and $| G{:}H | \leqslant | \mathbf{G}_1{:}\mathbf{H}_1 | \, | \mathbf{G}_2{:}\mathbf{H}_2 | < 2^{n-1}$.

***10.36.** First, let us suppose that H is completely reducible. Then the proof of 10.T.1 (as given, for example, in [1]) may be generalized to show that G is then completely reducible (see [45]). On the other hand, if G is completely reducible, then, by Problem 1.7, there is a normal subgroup N of finite index in G such that $N \subseteq H$. By Problem 10.8, N is completely reducible. Since N is of finite index in H, the first part of this problem shows that H is completely reducible.

***10.37.** See [45].

***10.38.** See [18], page 477. A generalization is proved in [16], page 496.

***10.39.** See [93].

***10.40.** This result is due to Goldberg [94], and may be proved by putting $\mathbf{u}$ into Jordan canonical form; the crucial step requires an analysis of the solutions of Pell's equation. An alternative proof comes from the theorem proved in Appendix B of [2], volume 2 together with the observation that an abelian subgroup of a free product of cyclic groups is cyclic ([2], Section 33). For a generalization see [95].

Solutions for Chapter 11—Representations and Characters

11.1. Let C_1, C_2, and C_3 be the conjugacy classes of S_3 containing 1, (12), and (123), respectively. Then the character table for S_3 is

	C_1	C_2	C_3
χ^1	1	1	1
χ^2	1	-1	1
χ^3	2	0	-1

11.2. Let C_1, C_2, C_3, and C_4 be the conjugacy classes of A_4 containing, 1, (12)(34), (123), and (132), respectively. Then the character table for A_4 is

	C_1	C_2	C_3	C_4
χ^1	1	1	1	1
χ^2	1	1	ω	ω^2
χ^3	1	1	ω^2	ω
χ^4	3	-1	0	0

where $\omega^3 = 1$ and $\omega \neq 1$. (For further details and examples see [16], Section 32.)

11.3. Let G have order g. From 11.T.4, G has g irreducible characters, and by 11.T.6 each of these characters has degree 1. Thus the irreducible representations of G are homomorphisms of G into $GL(1)$. Let x_i be a generator for C_i ($i = 1, 2, \ldots, m$), and let $\mathbf{R}$ be an irreducible representation of G. Then we must have $\mathbf{R}(x_i) = \omega_i$, where $\omega_i^{h_i} = 1$. Conversely, for any sequence $\omega_1, \omega_2, \ldots, \omega_m$ of complex numbers such that $\omega_i^{h_i} = 1$ (for $i = 1, 2, \ldots, m$), the mapping

$$x_1^{s_1} x_2^{s_2} \cdots x_m^{s_m} \rightarrow \omega_1^{s_1} \omega_2^{s_2} \cdots \omega_m^{s_m}$$

defines a homomorphism of G into $GL(1)$. Hence there is a one-to-one correspondence between the irreducible representations $\mathbf{R}$ of G and the finite sequences $\omega_1, \omega_2, \ldots, \omega_m$ of complex numbers with $\omega_i^{h_i} = 1$ for all i. (For further details see [1], Section 13.2.)

11.4. If we use the notation of Problem 5.11, the mapping

$$x \rightarrow \begin{bmatrix} 1 & 0 \\ \alpha & -1 \end{bmatrix} \quad \text{and} \quad y \rightarrow \begin{bmatrix} 0 & 1 \\ -1 & \alpha \end{bmatrix}$$

defines a faithful irreducible representation of G for any complex α such that $(\alpha + \sqrt{\alpha^2 - 4})/2$ is not a root of unity. All these representations are mutually inequivalent.

11.5. For both nonabelian groups of order 8 the character table has the form

$$\begin{matrix} 1 & 1 & 1 & 1 & 1 \\ 1 & 1 & -1 & -1 & 1 \\ 1 & 1 & -1 & 1 & -1 \\ 1 & 1 & 1 & -1 & -1 \\ 2 & -2 & 0 & 0 & 0 \end{matrix}$$

(For a related result see [96].)

11.6. Define two $n \times n$ permutation matrices $\mathbf{u} = [u_{ij}]$ and $\mathbf{v} = [v_{ij}]$ by: $u_{ij} = 1$ if and only if $j = i^u$; and $v_{ij} = 1$ if and only if $j = i^v$. Then $\mathbf{w}_1 = \mathbf{uw}$ and $\mathbf{w}_2 = \mathbf{wv}$. Since $\mathbf{w}_1 = \mathbf{w}_2$, $\mathbf{u} = \mathbf{wvw}^{-1}$, and so the trace of $\mathbf{u}$ equals the trace of $\mathbf{v}$. But tr $\mathbf{u}$ and tr $\mathbf{v}$ are the numbers of letters left fixed by u and v, respectively (see [10]).

11.7. We use the notation of 11.T.5. The mapping $\chi^i \rightarrow (\chi^i)^*$ ($i = 1, \ldots, k$) induces a permutation among the rows of the matrix $\mathbf{u}$

defined in 11.T.5. Similarly, the mapping $C_j \to C_{j^*}$ $(j = 1, \ldots, k)$ induces a permutation among the columns of $\mathbf{u}$. The two matrices which result from these permutations have as their (i,j)th entries $\sqrt{h_j/g}\,\bar{\chi}_j^i$ and $\sqrt{h_{j^*}/g}\,\chi_{j^*}^i$, respectively. But $h_j = h_{j^*}$ and $\bar{\chi}_j^i = \chi_{j^*}^i$ by 11.T.8, and so the result follows from Problem 11.6.

11.8. Let C_i be a conjugacy class of G. The condition $C_i = C_{i^*}$ holds if and only if each $x \in C_i$ is conjugate to x^{-1}; that is, $x = u^{-1}x^{-1}u$ for some $u \in G$. Then $x = u^{-s}x^{-1}u^s$ for each odd integer s, and so, taking s to be the order of u, $x = x^{-1}$. Since x has odd order, this implies that $x = 1$. Thus the only self-inverse class C_i is the class $\{1\}$. From Problem 11.7 we conclude that G has only one real irreducible character, namely, 1_G.

11.9. From Problem 11.8 we know that the distinct irreducible characters of G may be written: $\chi^0 = 1_G$, χ^1, $(\chi^1)^*, \ldots, \chi^l$, $(\chi^l)^*$ where $k = 2l + 1$. The degrees of these characters are all odd, by 11.T.6, and so we may write $2c_i + 1$ for the degree of χ^i [and $(\chi^i)^*$]. Then, from 11.T.6,

$$n = 1 + 2\sum_{i=1}^{l}(2c_i + 1)^2 = 1 + 2l + 8\sum_{i=1}^{l} c_i(c_i + 1).$$

Since $c_i(c_i + 1)$ is even for all integers c_i, n is congruent (mod 16) to $1 + 2l = k$ (see [18], page 294).

11.10. If G has a faithful irreducible representation, then $Z(G)$ is cyclic by Problem 10.16.

Conversely, if $Z(G)$ is cyclic, then it contains a unique subgroup N of order p. From Problem 6.13, it follows that N is the only minimal normal subgroup of G. Let χ be the character corresponding to some faithful representation of G (for example, the regular representation), and suppose that $\chi = \sum_a c_a\chi^a$, where the χ^a are the irreducible characters of G. Choose $x \neq 1$ in N. Since x does not lie in the kernel of χ, there is some irreducible character χ^a such that x is not in the kernel of χ^a. Since the kernel of χ^a does not contain N, it follows from the choice of N that the kernel of χ^a is 1; that is, χ^a is a faithful character. (For a more general result, see [97].)

11.11. Let χ be the character corresponding to the given representation $\mathbf{R}$. Let d be the degree of any irreducible constituent of χ. By 11.T.6, d divides $|G|$. Since d is less than the smallest prime dividing $|G|$, $d = 1$. Thus all the irreducible constituents of $\mathbf{R}(G)$ have degree 1. By 10.T.1, $\mathbf{R}(G)$ is completely reducible, so $\mathbf{R}(G)$ is similar to a group of diagonal matrices. Hence G, which is isomorphic to $\mathbf{R}(G)$, is abelian.

11.12. If $p^2 \nmid n$, then the Sylow p-group is cyclic. So let us consider the case $p^2 \mid n$. Then the number k of conjugacy classes of G is at least $n/p^2 + 1$. Let $d_1 \leqslant d_2 \leqslant \cdots \leqslant d_k$ be the degrees of the irreducible characters of G, where $d_1 = 1$ corresponds to the identity character 1_G. By 11.T.6,

$$n = \sum_{a=1}^{k} d_a^2 \geqslant 1 + d_2^2 n p^{-2}.$$

Thus $d_2^2 \leqslant [p^2(n-1)]/n$, and so $d_2 < p$. Hence G has a representation $\mathbf{R}$ of degree $d_2 < p$. Moreover $\mathbf{R}$ is faithful because G is simple, and so the restriction of $\mathbf{R}$ to a Sylow p-group P of G gives a faithful representation of P of degree d_2. From Problem 11.11, we conclude that P is abelian.

11.13. Let $r_1, r_2, \ldots, r_n$ be a set of left coset representatives for H in G. Then, for each $u \in H$ and $x \in G$, we have $\breve{\chi}(u^{-1}r_i^{-1}xr_iu) = \breve{\chi}(r_i^{-1}xr_i)$, by 11.T.2. Therefore, for all $x \in G$,

$$\begin{aligned}\sum_{y\in G} \breve{\chi}(y^{-1}xy) &= \sum_{i=1}^{n} \sum_{u\in H} \breve{\chi}(u^{-1}r_i^{-1}xr_iu) = \sum_{i=1}^{n} |\,H\,|\, \chi(r_i^{-1}xr_i)\\ &= |\,H\,|\, \chi^G(x)\end{aligned}$$

as asserted.

11.14. Let g and h be the orders of G and H, respectively. Using Problem 11.13, we have

$$\begin{aligned}\langle \psi^G, \chi\rangle &= \frac{1}{gh} \sum_{y\in G} \sum_{x\in G} \breve{\psi}(y^{-1}xy)\overline{\chi(x)}\\ &= \frac{1}{gh} \sum_{y\in G} \sum_{x\in G} \breve{\psi}(y^{-1}xy)\overline{\chi(y^{-1}xy)} \qquad \text{by 11.T.2}\\ &= \frac{1}{gh} \sum_{y\in G} \sum_{z\in G} \breve{\psi}(z)\overline{\chi(z)} \qquad \text{putting } z = y^{-1}xy\\ &= \frac{1}{h} \sum_{z\in H} \psi(z)\overline{\chi(z)} = \langle \psi, \chi|H\rangle.\end{aligned}$$

11.15. By Problem 1.6 G has five conjugacy classes C_1, C_2, C_3, C_4, and C_5 containing 1, a, b, b^2, and b^3, respectively. The character table for G is

	C_1	C_2	C_3	C_4	C_5
χ^1	1	1	1	1	1
χ^2	1	-1	1	1	1
χ^3	2	0	$\xi + \xi^6$	$\xi^2 + \xi^5$	$\xi^3 + \xi^4$
χ^4	2	0	$\xi^2 + \xi^5$	$\xi^3 + \xi^4$	$\xi + \xi^6$
χ^5	2	0	$\xi^3 + \xi^4$	$\xi + \xi^6$	$\xi^2 + \xi^5$

where $\xi^7 = 1$, $\xi \neq 1$. The last three characters are the characters of G induced from irreducible characters of H (see [16], page 339).

11.16. Let C_1, C_2, C_3, C_4, and C_5 be the conjugacy classes of A_5 containing 1, (12)(34), (123), (12345), and (13245), respectively. Then the character table for A_5 is

	C_1	C_2	C_3	C_4	C_5
χ^1	1	1	1	1	1
χ^2	3	-1	0	$-\zeta - \zeta^4$	$-\zeta^2 - \zeta^3$
χ^3	3	-1	0	$-\zeta^2 - \zeta^3$	$-\zeta - \zeta^4$
χ^4	4	0	1	-1	-1
χ^5	5	1	-1	0	0

where $\zeta^5 = 1$, $\zeta \neq 1$.

11.17. Let ψ be a character of H occurring as an irreducible constituent of $\chi|H$. By Problem 11.14, χ occurs as a constituent of ψ^G. Hence $\chi(1) \leqslant \psi^G(1) = m\psi(1)$. If H is abelian, then ψ has degree 1, by Problem 10.1, and so $\chi(1) \leqslant m$.

11.18. First, let us suppose that G is transitive and that H is a stabilizer for G. If H fixes the letter i, then we can choose $r_1, r_2, \ldots, r_n$ as a set of left coset representatives for H in G such that r_j maps j to i. Then $r_jHr_j^{-1}$ is the subgroup of G fixing j (see Problem 2.11), and so an element x in G fixes j if and only if $r_j^{-1}xr_j \in H$. Thus $1_H^G(x) = \sum_{j=1}^n \breve{1}_H(r_j^{-1}xr_j)$ equals the number of letters fixed by x. Hence $1_H^G = \pi$ as asserted.

In general, if $\Omega_1, \Omega_2, \ldots, \Omega_s$ are the orbits for G, then $G|\Omega_i$ is a transitive permutation group, and so the permutation character for $G|\Omega_i$ is $1_{H_i}^G$, where H_i is a stabilizer of $G|\Omega_i$. The permutation character π of G is the sum of the characters of the groups $G|\Omega_i$, and so the result follows.

11.19. First, we suppose that G has s orbits. Then, using the notation of Problem 11.18, we have

$$\begin{aligned}\langle \pi, 1_G\rangle &= \sum_{i=1}^{s} \langle 1_{H_i}^G, 1_G\rangle = \sum_{i=1}^{s} \langle 1_{H_i}, 1_G|H_i\rangle \qquad \text{by Problem 11.14}\\ &= s \qquad \text{because } 1_G|H_i = 1_{H_i}.\end{aligned}$$

Next, let us suppose that G is transitive and that H is a stabilizer of G. Then, by Problem 11.18, $\pi = 1_H^G$. Hence $\langle \pi,\pi\rangle = \langle 1_H, \pi|H\rangle$, by Problem 11.14. Since $\pi|H$ is the permutation character of H, the first part of the proof shows that H has $\langle \pi,\pi\rangle$ orbits.

11.20. Let H be any stabilizer of G. Then, by Problem 11.18, $\pi = 1_H^G$. Hence, by Problem 11.14, $\langle \pi, \chi \rangle = \langle 1_H, \chi|H \rangle \leqslant \deg (\chi|H) = \deg \chi$ (using 11.T.7). We have equality precisely when $\chi|H = (\deg \chi)\, 1_H$, that is, when $H \subseteq \ker \chi$. Thus equality holds if and only if the normal subgroup $\ker \chi$ contains all the (conjugate) stabilizers of G.

11.21. Let $\mathbf{R}$ be an irreducible representation for G whose kernel contains N. Then $\mathbf{R}_1$ defined by $\mathbf{R}_1(Nx) = \mathbf{R}(x)$ $(Nx \in G/N)$ is well defined and is clearly a representation of G/N. Moreover $\mathbf{R}_1$ is irreducible because $\mathbf{R}$ is irreducible. Conversely, if $\mathbf{S}_1$ is an irreducible representation of G/N, then $\mathbf{S}$ defined by $\mathbf{S}(x) = \mathbf{S}_1(Nx)$ $(x \in G)$ is an irreducible representation for G. Furthermore, $\mathbf{R}_1$ is equivalent to $\mathbf{S}_1$ if and only if $\mathbf{R}$ is equivalent to $\mathbf{S}$. The given result on characters now follows.

11.22. Let $\alpha^1, \alpha^2, \ldots, \alpha^s$ be the irreducible characters of G which contain N in their kernels, and let $\chi = \chi^1, \chi^2, \ldots, \chi^r$ be the other irreducible characters of G. Then, with the notation of Problem 11.21, 11.T.5 gives

$$\sum_{i=1}^{s} | \alpha^i(x) |^2 = \sum_{i=1}^{s} | \beta^i(Nx) |^2 = | C(Nx; G/N) |,$$

and

$$\sum_{i=1}^{s} | \alpha^i(x) |^2 + \sum_{j=1}^{r} | \chi^j(x) |^2 = | C(x; G) |.$$

Because $C(x;N) = 1$, at most one element of any coset Nu in G lies in $C(x;G)$. Moreover, if one element of Nu lies in $C(x;G)$, then Nu lies in $C(Nx;G/N)$. Hence, $| C(x;G) | \leqslant | C(Nx;G/N) |$. Thus the equations above imply that $\sum_{j=1}^{r} | \chi^j(x) |^2 \leqslant 0$. In particular, $\chi(x) = 0$ (see [10]).

11.23. Using the notation of Equation (11.1), we have

$$\operatorname{tr} (\mathbf{x} \otimes \mathbf{y}) = \sum_{i=1}^{n} \operatorname{tr} (\mathbf{x}\eta_{ii}) = \sum_{i=1}^{n} (\operatorname{tr} \mathbf{x})\eta_{ii} = (\operatorname{tr} \mathbf{x})(\operatorname{tr} \mathbf{y}).$$

To prove the second part, choose a matrix $\mathbf{u}$ such that

$$\mathbf{u}^{-1}\mathbf{y}\mathbf{u} = \begin{bmatrix} \gamma_{11} & \gamma_{12} & \cdots & \gamma_{1n} \\ & \gamma_{22} & \cdots & \gamma_{2n} \\ & & \ddots & \vdots \\ & 0 & & \vdots \\ & & & \gamma_{nn} \end{bmatrix}.$$

Then $\det \mathbf{y} = \gamma_{11}\gamma_{22}\cdots\gamma_{nn}$. Also, we have

$$(\mathbf{1}\otimes\mathbf{u})^{-1}(\mathbf{x}\otimes\mathbf{y})(\mathbf{1}\otimes\mathbf{u}) = \mathbf{x}\otimes(\mathbf{u}^{-1}\mathbf{y}\mathbf{u}) = \begin{bmatrix} \mathbf{x}\gamma_{11} & \mathbf{x}\gamma_{12} & \cdots & \mathbf{x}\gamma_{1n} \\ & \mathbf{x}\gamma_{22} & \cdots & \mathbf{x}\gamma_{2n} \\ & & \ddots & \vdots \\ & 0 & & \vdots \\ & & & \mathbf{x}\gamma_{nn} \end{bmatrix}.$$

Hence

$$\det(\mathbf{x}\otimes\mathbf{y}) = \prod_{i=1}^{n} \det(\mathbf{x}\gamma_{ii}) = \prod_{i=1}^{n} \gamma_{ii}^{m}(\det \mathbf{x})^{n} = (\det \mathbf{x})^{n}(\det \mathbf{y})^{m}.$$

11.24. Let $\mathbf{R}$ and $\mathbf{S}$ be representations of G corresponding to the characters χ and ψ, respectively. Then the mapping $x \to \mathbf{R}(x)\otimes\mathbf{S}(x)$ $(x\in G)$ is a representation for G, from Equation (11.2). By Problem 11.23, the character of this representation is $\chi\psi$.

11.25. Let $\mathcal{V}$ be a vector space of dimension n over $\mathcal{C}$. From the correspondence between $GL(\mathcal{V})$ and $GL(n)$ we can find a homomorphism R of $G\times H$ into $GL(\mathcal{V})$ corresponding to the matrix representation $\mathbf{R}$. Let $\mathcal{W}$ be a minimal invariant subspace for $R(G)$. Then, by Problem 10.7,

$$\mathcal{V} = \mathcal{W}R(u_1)\dotplus\cdots\dotplus\mathcal{W}R(u_k),$$

for some elements u_i in H. [Take $R(G\times H)$ for G and $R(G)$ for H in Problem 10.7.] If $v_1,\ldots,v_m$ is a basis for $\mathcal{W}$, then $v_1R(u_j),\ldots,v_mR(u_j)$ is a basis for $\mathcal{W}R(u_j)$ for each j. Since u_j centralizes G, for all $x\in G$, we have

$$v_iR(u_j)R(x) = v_iR(x)R(u_j) \qquad (i = 1, 2, \ldots, m).$$

Thus the matrix $\mathbf{S}(x)$, which corresponds to the action of x on the subspace $\mathcal{W}R(u_j)$ relative to the given basis, is the same for each j. Therefore, over the basis

$$v_1R(u_1), v_2R(u_1),\ldots, v_mR(u_1),\ldots, v_1R(u_k), v_2R(u_k),\ldots, v_mR(u_k)$$

of $\mathcal{V}$ we obtain a representation $\mathbf{R}_1$ of $G\times H$ which is equivalent to $\mathbf{R}$ and such that

$$\mathbf{R}_1(x) = \begin{bmatrix} \mathbf{S}(x) & \mathbf{0} & \cdots & \mathbf{0} \\ \mathbf{0} & \mathbf{S}(x) & \cdots & \mathbf{0} \\ \vdots & & & \vdots \\ \mathbf{0} & \mathbf{0} & \cdots & \mathbf{S}(x) \end{bmatrix} \qquad \text{for all } x\in G.$$

The representation $\mathbf{S}$ for G is irreducible because $\mathcal{W}$ is a minimal invariant subspace for $R(G)$. Let us suppose that

$$\mathbf{R}_1(y) = \begin{bmatrix} \mathbf{R}_{11}(y) & \mathbf{R}_{12}(y) & \cdots & \mathbf{R}_{1k}(y) \\ \mathbf{R}_{21}(y) & \mathbf{R}_{22}(y) & \cdots & \mathbf{R}_{2k}(y) \\ \cdot & & & \cdot \\ \cdot & & & \cdot \\ \cdot & & & \cdot \\ \mathbf{R}_{k1}(y) & \mathbf{R}_{k2}(y) & \cdots & \mathbf{R}_{kk}(y) \end{bmatrix} \qquad \text{for } y \in H,$$

where the matrix is partitioned into $m \times m$ blocks. Since each $y \in H$ commutes with each $x \in G$, we obtain $\mathbf{R}_1(x)\mathbf{R}_1(y) = \mathbf{R}_1(y)\mathbf{R}_1(x)$; that is, $\mathbf{S}(x)\mathbf{R}_{ij}(y) = \mathbf{R}_{ij}(y)\mathbf{S}(x)$ for all $x \in G$. Since $\mathbf{S}$ is irreducible, $\mathbf{R}_{ij}(y)$ is a scalar matrix $\eta_{ij}(y)\mathbf{1}$, by Problem 10.1. Hence $\mathbf{R}_1(y) = \mathbf{1} \otimes \mathbf{T}(y)$, where $\mathbf{T}(y) = [\eta_{ij}(y)]$ for each $y \in H$. Thus we see that $\mathbf{T}$ is a representation of H, and

$$\begin{aligned} \mathbf{R}_1(xy) &= (\mathbf{S}(x) \otimes \mathbf{1})(\mathbf{1} \otimes \mathbf{T}(y)) \\ &= \mathbf{S}(x) \otimes \mathbf{T}(y) \qquad \text{for all } x \in G \text{ and } y \in H. \end{aligned}$$

We have already noted that $\mathbf{S}$ is irreducible, and it is readily seen that $\mathbf{T}$ must also be irreducible. Thus the first part of the result is proved. The result in terms of characters then follows immediately.

11.26. Let $\mathbf{R}$ be a representation for G having the character χ. By 11.T.8, for each $x \in G$, $\chi(x)$ is a sum of d roots of unity, say, $\omega_1 + \omega_2 + \cdots + \omega_d$ Hence $|\chi(x)| \leqslant |\omega_1| + |\omega_2| + \cdots + |\omega_d| = d$. We have equality if and only if all the complex numbers ω_i have the same argument. Since all the ω_i have the same absolute value (namely, 1), we conclude that $|\chi(x)| = d$ if and only if $\omega_1 = \omega_2 = \cdots = \omega_d = \omega$, say. Since $\mathbf{R}(x)$ is similar to a diagonal matrix, by 11.T.8, $|\chi(x)| = d$ if and only if $\mathbf{R}(x) = \omega\mathbf{1}$. Thus (a) now follows from Problem 10.1. In case (b), ω must be 1, and so $\mathbf{R}(x) = \mathbf{1}$. Hence $x \in K$.

11.27. Define $A_i = \{x \in G \mid \chi(x) = \alpha_i\}$. From Problem 11.26, we may take $\alpha_1 = d$, and then $A_1 = \{1\}$ because χ is faithful. Since $\chi^{(s)}(x) = \alpha_i^s$ for all $x \in A_i$, therefore

$$\langle \chi^{(s)}, \psi \rangle = \frac{1}{g} \sum_{i=1}^{r} \alpha_i^s \sum_{x \in A_i} \overline{\psi(x)} = \frac{1}{g} \sum_{i=1}^{r} \alpha_i^s u_i \qquad (s = 0, 1, \ldots, r-1).$$

This is a set of r linear equations in the r quantities $u_i = \sum_{x \in A_i} \overline{\psi(x)}$. The Vandermonde determinant $\det[\alpha_i^s] = \prod_{i<j}(\alpha_i - \alpha_j)$ is not zero. Hence the equations have a unique solution for the u_i. Since $u_1 = d$, this solution is not trivial, and so, for some s with $0 \leqslant s < r$, $\langle \chi^{(s)}, \psi \rangle \neq 0$, Thus ψ is a constituent of $\chi^{(s)}$.

Finally, every irreducible character of G is a constituent of the character $\sum_{s=0}^{r-1} \chi^{(s)}$. Therefore the sum of the degrees of the irreducible characters of G is at most $\sum_{s=0}^{r-1} d^s = (d^r - 1)/(d - 1)$ (see [98]).

11.28. Let $r_1, r_2, \ldots, r_n$ be a set of left coset representatives for H in G. Put $\theta = (\chi|H)\psi$. Then $\breve{\theta} = \chi\breve{\psi}$. Hence, for each $x \in G$,

$$\begin{aligned}\theta^G(x) &= \sum_{i=1}^{n} \chi(r_i^{-1}xr_i)\breve{\psi}(r_i^{-1}xr_i) \\ &= \chi(x) \sum_{i=1}^{n} \breve{\psi}(r_i^{-1}xr_i) \qquad \text{by 11.T.2} \\ &= \chi(x)\psi^G(x)\end{aligned}$$

as asserted (see [99]).

11.29. Let $r_1, r_2, \ldots, r_n$ be a set of coset representatives for N in G. Then, for each $x \in N$,

$$\phi^G(x) = \sum_{i=1}^{n} \breve{\phi}(r_i^{-1}xr_i) = \sum_{i=1}^{n} \chi(r_i^{-1}xr_i) = n\chi(x)$$

from 11.T.2. Therefore, by Problem 11.14, $\langle\phi^G,\phi^G\rangle = \langle\chi|N,\phi^G|N\rangle = n\langle\chi|N,\chi|N\rangle = n$ because $\chi|N$ is irreducible.

11.30. From Problem 11.28, $(\chi|N)^G = \{(\chi|N)1_N\}^G = \chi 1_N^G$. Write $1_N^G = \sum_a c_a\psi^a$, where ψ^a are all the irreducible characters of G. Then, using Problem 11.29, we have

$$\begin{aligned}n = \langle\chi 1_N^G, \chi 1_N^G\rangle &= \langle\sum_a c_a\chi\psi^a, \sum_a c_a\chi\psi^a\rangle \\ &= \sum_{a,b} c_a c_b\langle\chi\psi^a,\chi\psi^b\rangle \\ &\geqslant \sum_a c_a^2 \langle\chi\psi^a, \chi\psi^a\rangle \qquad \text{since all } c_a \geqslant 0 \\ &\geqslant \sum_a c_a^2 \qquad \text{since } \langle\chi\psi^a,\chi\psi^a\rangle \geqslant 1 \\ &= \langle 1_N^G, 1_N^G\rangle = n \qquad \text{because} \quad 1_G|N = 1_N\end{aligned}$$

(using Problem 11.29).

Thus, we must have equality throughout and, in particular, $\langle\chi\psi^a,\chi\psi^a\rangle = 1$ whenever $c_a \neq 0$. But, using 11.T.7 and Problem 11.14, we find $c_a = \langle 1_N^G, \psi^a\rangle = \langle 1_N, \psi^a|N\rangle = (\deg \psi^a)\langle 1_N, 1_N\rangle \neq 0$ whenever N is contained in the kernel of ψ^a. Thus, in this case, $\langle\chi\psi^a, \chi\psi^a\rangle = 1$; that is, $\chi\psi^a$ is irreducible (see [100]).

11.31. We first note that, if K is a subgroup of a finite group H, and K has an irreducible character χ of degree d, then H has an irreducible character of degree $\geqslant d$, namely, any constituent of χ^H (by Problem 11.14). Thus the asserted result is a corollary of the following result:

Let N be a normal subgroup of a finite group H such that H/N is nonabelian. If N has an irreducible character of degree d, then H has an irreducible character of degree $\geqslant 2d$.

We prove this latter result. Let ψ be a character of degree d for N, and let χ be an irreducible constituent of the induced character ψ^H. From Problem 10.8, $\chi|N$ is a sum of t irreducible characters of equal degrees. By Problem 11.14, one of these characters is ψ, and so the degree of χ equals td. If $t \geqslant 2$, we are finished, so let us consider the case $t = 1$; that is, $\chi|N$ is irreducible and of degree d. Since H/N is not abelian, it has an irreducible character of degree at least 2, using 11.T.6. Hence, using Problem 11.21, we may find an irreducible character α of H with $N \subseteq \ker \alpha$ and the degree of α at least 2. Then, by Problem 11.30, $\chi\alpha$ is an irreducible character of H, and $\deg \chi\alpha = (\deg \chi)(\deg \alpha) \geqslant 2d$ (see [100]).

11.32. Let χ be the character of the given representation $\mathbf{R}$. If $z \neq 1$ is an element of $Z(G)$, then $\mathbf{R}(z) = \zeta\mathbf{1}$ for some scalar $\zeta \neq 1$, by Problem 10.1. Therefore, for each $x \in G$, $\chi(xz) = \operatorname{tr} \mathbf{R}(xz) = \zeta \operatorname{tr} \mathbf{R}(x) = \zeta \cdot \chi(x)$. If $x \notin Z(G)$, then we may choose $y \in G$ such that $yx \neq xy$. By hypothesis, $z = [x,y] \neq 1$ is in $Z(G)$, and so $\chi(x) = \chi(y^{-1}xy) = \chi(xz) = \zeta \cdot \chi(x)$, and therefore $\chi(x) = 0$. Since $|\chi(z)| = n$ for all $z \in Z(G)$ and is 0 otherwise, we have

$$1 = \langle \chi,\chi \rangle = \frac{1}{g} \sum_{z \in Z(G)} |\chi(z)|^2 = n^2 |Z(G)|/g.$$

Hence $|G:Z(G)| = n^2$ (see [100]).

11.33. Let $\mathcal{V}$ be the vector space over $\mathcal{C}$ of all $1 \times n$ matrices with entries in $\mathcal{C}$, and let G correspond to the group $\mathbf{G}$ under the isomorphism between $GL(\mathcal{V})$ and $GL(n)$ arising when we choose the standard basis $e_i = (\delta_{i1}\delta_{i2} \cdots \delta_{in})$ $(i = 1, 2, \ldots, n)$ for $\mathcal{V}$. (δ_{ij} is the Kronecker delta.) Let $\mathcal{V}_k$ denote the subspace of $\mathcal{V}$ consisting of the kth rows of all matrices in $\mathfrak{M}$ $(k = 1, 2, \ldots, n)$. Then $\mathcal{V}_k$ is clearly an invariant subspace of G and, since G is irreducible (because $\mathbf{G}$ is), therefore $\mathcal{V}_k = \mathcal{V}$ or 0. Since $\mathfrak{M} \neq 0$, $\mathcal{V}_l = \mathcal{V}$ for some l. Thus the dimension m of $\mathfrak{M}$ is at least as great as the dimension n of $\mathcal{V}$.

Suppose that $m > n$. Then choose a basis $\mathbf{u}_1, \ldots, \mathbf{u}_m$ of $\mathfrak{M}$ such that the lth row of $\mathbf{u}_i$ is the standard basis vector e_i if $i \leqslant n$, and is $(0\,0 \cdots 0)$

for $i > n$. This is possible because $\mathcal{V}_l = \mathcal{V}$. Then the matrices in $\mathfrak{M}$ with their lth rows equal to $(0\ 0 \cdots 0)$ form a nontrivial invariant subspace $\mathfrak{N}$ for G; that is, $\mathfrak{N}\mathbf{G} = \mathfrak{N}$ contrary to the choice of $\mathfrak{M}$. We conclude that $m = n$.

11.34. We use the notation of the proof of Problem 11.33. Let us suppose that $\mathbf{G}^\perp \neq \mathbf{0}$. Then, since $\mathbf{G}^\perp$ is an invariant subspace for G, it contains a minimal invariant subspace $\mathfrak{M}$. As in the proof of Problem 11.33, we may choose a basis $\mathbf{u}_1, \ldots, \mathbf{u}_n$ for $\mathfrak{M}$ so that the lth row of $\mathbf{u}_i$ is e_i $(i = 1, \ldots, n)$.

Let $(\sigma_{i1}\sigma_{i2} \cdots \sigma_{in})$ denote the sth row of $\mathbf{u}_i$. Then the vectors $v_i = (\sigma_{i1} - \alpha\delta_{i1}\ \sigma_{i2} - \alpha\delta_{i2} \cdots \sigma_{in} - \alpha\delta_{in})$ $(i = 1, \ldots, n)$ span an invariant subspace for G for any $\alpha \in \mathcal{C}$. Since $\mathcal{C}$ is algebraically closed, we may choose $\alpha = \alpha_s$ so that $\det[\sigma_{ij} - \alpha_s\delta_{ij}] = 0$. Then the vectors v_i are linearly dependent and so span the zero subspace (because G is irreducible); that is, $v_i = 0$ for all i. Thus for each s there exists $\alpha_s \in \mathcal{C}$ such that $\sigma_{ij} = \alpha_s\delta_{ij}$ for all i,j. Hence

$$\mathbf{u}_i = \begin{bmatrix} \alpha_1 \\ \alpha_2 \\ \cdot \\ \cdot \\ \cdot \\ \alpha_n \end{bmatrix} [\delta_{i1}\quad \delta_{i2}\quad \cdots\quad \delta_{in}] \qquad (\text{for } i = 1, \ldots, n).$$

Finally, by the definition of $\mathbf{G}^\perp$, we have $\alpha_i = \operatorname{tr}(\mathbf{u}_i \cdot \mathbf{1}) = 0$. Thus $\mathfrak{M} = \mathbf{0}$ contrary to our assumption. Hence $\mathbf{G}^\perp = \mathbf{0}$.

We now show that the dimension of $\mathcal{C}\mathbf{G}$ is n^2. Let us choose a basis $\mathbf{x}_1, \mathbf{x}_2, \ldots, \mathbf{x}_s$ $(s \leqslant n^2)$ for $\mathcal{C}\mathbf{G}$ consisting of elements of $\mathbf{G}$. Then $\operatorname{tr}(\mathbf{y}\mathbf{x}_r) = 0$ (for $r = 1, 2, \ldots, s$) implies that $\mathbf{y} \in \mathbf{G}^\perp$, and so $\mathbf{y} = \mathbf{0}$. Put $\mathbf{y} = [\eta_{ij}]$ and $\mathbf{x}_r = [\xi_{rij}]$. Then the s simultaneous equations

$$\operatorname{tr}(\mathbf{y}\mathbf{x}_r) = \sum_{i,j} \eta_{ji}\xi_{rij} = 0 \qquad (r = 1, 2, \ldots, s)$$

in the n^2 quantities η_{ji} have only the trivial solution: $\eta_{ji} = 0$ for all i,j. This implies that $s \geqslant n^2$, and so $s = n^2$ as asserted. (This theorem is due to Burnside. For another proof see [16], page 182.)

11.35. Choose $\mathbf{y}_1, \mathbf{y}_2, \ldots, \mathbf{v}_s$ in $\mathbf{G}$ as a basis for the subspace $\mathcal{C}\mathbf{G}$ of $\mathfrak{L}$. Then $s = n^2$, by Problem 11.34. For each $\mathbf{x} \in \mathbf{G}$, we have a sequence

$$\operatorname{tr}(\mathbf{x}\mathbf{y}_1),\ \operatorname{tr}(\mathbf{x}\mathbf{y}_2),\ \ldots,\ \operatorname{tr}(\mathbf{x}\mathbf{y}_s).$$

Two different elements $\mathbf{x}$ and $\mathbf{x}'$ in $\mathbf{G}$ cannot have the same sequence because $\operatorname{tr}(\mathbf{x}\mathbf{y}_i) = \operatorname{tr}(\mathbf{x}'\mathbf{y}_i)$ $(i = 1, 2, \ldots, s)$ implies $\operatorname{tr}(\mathbf{x}\mathbf{y}) = \operatorname{tr}(\mathbf{x}'\mathbf{y})$ for all $\mathbf{y} \in \mathbf{G}$, and so $\mathbf{x} - \mathbf{x}' = \mathbf{0}$ by Problem 11.34. Thus G has no more elements than there are possible sequences of trace values, namely, $m^s = m^{n^2}$.

11.36. Let $\mathbf{S}$ be a faithful representation of degree n for G. Then the character corresponding to $\mathbf{S}$ takes only a finite number of values, by 11.T.2. Thus, in the case that $\mathbf{S}$ is irreducible, the result follows immediately from Problem 11.35. Therefore, consider the case where $\mathbf{S}$ is reducible.

If $\mathbf{S}$ is reducible, then there is a representation $\mathbf{S}_0$ equivalent to $\mathbf{S}$ such that

$$\mathbf{S}_0(x) = \begin{bmatrix} \mathbf{S}_1(x) & \mathbf{0} \\ \mathbf{S}_3(x) & \mathbf{S}_2(x) \end{bmatrix} \qquad \text{for all } x \in G,$$

where $\mathbf{S}_1$ and $\mathbf{S}_2$ are representations for G of degrees r and $n - r$, respectively, for some r with $0 < r < n$. Let N_1 and N_2 be the corresponding kernels so that $N_1 \cap N_2 = \{x \in G \mid \mathbf{S}_1(x) = \mathbf{1} \text{ and } \mathbf{S}_2(x) = \mathbf{1}\}$. Since G/N_1 and G/N_2 are groups with a finite number of conjugacy classes, we may apply induction on the degree of the representations to deduce that G/N_1 and G/N_2 are finite groups. Hence $|G{:}N_1 \cap N_2| = h$, say, is finite. The group $N_1 \cap N_2$ is abelian, and so $C(x;G)$ contains $N_1 \cap N_2$ for each $x \in N_1 \cap N_2$. Thus each element of $N_1 \cap N_2$ lies in a conjugacy class of G consisting of at most h elements. But G has a finite number of conjugacy classes, so $|N_1 \cap N_2|$ is finite. Hence $|G| = h\,|N_1 \cap N_2|$ is finite.

11.37. An $n \times n$ matrix $\mathbf{x}$ which satisfies $\mathbf{x}^m = \mathbf{1}$ has each of its eigenvalues equal to an mth root of unity. Thus $\operatorname{tr}\mathbf{x}$ (as the sum of these eigenvalues) can have any one of m^n different values. Thus $\operatorname{tr}\mathbf{x}$ $(\mathbf{x} \in \mathbf{G})$ takes at most m^n different values.

We now proceed by induction on the degree n of $\mathbf{G}$. If $\mathbf{G}$ is irreducible, then, by Problem 11.35, $|\mathbf{G}| \leqslant (m^n)^{n^2}$ as asserted. Therefore we consider the case when $\mathbf{G}$ is reducible. In this case, $\mathbf{G}$ is similar to a group $\mathbf{G}_0$ whose matrices have the form

$$(*) \qquad \mathbf{x} = \begin{bmatrix} \mathbf{x}_1 & \mathbf{0} \\ \mathbf{x}_3 & \mathbf{x}_2 \end{bmatrix}$$

where $\mathbf{x}_1$ and $\mathbf{x}_2$ are square blocks of degrees r and $n - r$, respectively, and $0 < r < n$. Since $\mathbf{x}_i^m = \mathbf{1}$, the group $\mathbf{G}_i = \{\mathbf{x}_i \text{ in } (*) \mid \mathbf{x} \in \mathbf{G}_0\}$ satisfies the condition of the induction hypothesis (for $i = 1, 2$). Therefore $|\mathbf{G}_1| \leqslant m^{r^3}$ and $|\mathbf{G}_2| \leqslant m^{(n-r)^3}$. Since

$$\begin{bmatrix} \mathbf{1}_r & \mathbf{0} \\ \mathbf{a} & \mathbf{1}_{n-r} \end{bmatrix}^m = \begin{bmatrix} \mathbf{1}_r & \mathbf{0} \\ m\mathbf{a} & \mathbf{1}_{n-r} \end{bmatrix},$$

the only matrix $\mathbf{x} \in \mathbf{G}_0$ for which $\mathbf{x}_1 = \mathbf{1}$ and $\mathbf{x}_2 = \mathbf{1}$ in (*) is the unit matrix. Thus each matrix in $\mathbf{G}_0$ is completely determined by the values $\mathbf{x}_1$ and $\mathbf{x}_2$ in (*), and so $|\mathbf{G}| = |\mathbf{G}_0| \leqslant |\mathbf{G}_1||\mathbf{G}_2| \leqslant m^{r^3+(n-r)^3} < m^{n^3}$ as asserted. (For the results of Problems 11.36 and 11.37 see [101]. See also [16], Section 36.)

11.38. We know that $\mathbf{G}$ is similar to a group $\mathbf{G}_0$ whose matrices have the form

$$(**) \qquad \mathbf{x} = \begin{bmatrix} \mathbf{x}_1 & & & \mathbf{0} \\ \mathbf{x}_{21} & \mathbf{x}_2 & & \\ \vdots & & \ddots & \\ \mathbf{x}_{k1} & \mathbf{x}_{k2} & \cdots & \mathbf{x}_k \end{bmatrix}$$

where the groups $\mathbf{G}_i = \{\mathbf{x}_i \text{ in } (**) \mid \mathbf{x} \in \mathbf{G}_0\}$ of degree n_i, say, are all irreducible. From the hypothesis, all the eigenvalues of $\mathbf{x}_i$ equal 1 for all $\mathbf{x} \in \mathbf{G}_0$. Thus tr $\mathbf{x}_i = n_i$ for all $\mathbf{x}_i \in \mathbf{G}_i$, and so $|\mathbf{G}_i| = 1$ by Problem 11.35. Hence each $n_i = 1$, and so each $\mathbf{x}_i = \mathbf{1}$ and each $\mathbf{x} \in \mathbf{G}_0$ is in $STL(n)$. (See [102].)

11.39. Let $\mathbf{R}$ be a representation of G corresponding to the character χ. Let $u_1, \ldots, u_m$ be a set of coset representatives for $Z(G)$ in G. By Problem 10.1, $\mathbf{R}(z)$ is a scalar matrix for each $z \in Z(G)$. Therefore the matrix $\mathbf{R}(zu_i)$ is a scalar multiple of $\mathbf{R}(u_i)$, and so the matrices $\mathbf{R}(u_1), \ldots, \mathbf{R}(u_m)$ span the vector space $\mathcal{C}\mathbf{R}(G)$. Thus, using Problem 11.34, we have $(\deg \chi)^2 = \chi(1)^2 =$ dimension of $\mathcal{C}\mathbf{R}(G) \leqslant m = |G:Z(G)|$.

***11.40.** This result is due to K. Taketa, and his proof is given in [5], page 366, and [16], page 356. The converse is false, and an example due to Itô showing this is quoted in [5]. However, E. Dade has recently proved that every finite solvable group G may be embedded as a subgroup in a finite solvable group G_1 which has the property that every representation is equivalent to a representation in terms of monomial matrices.

References

1. M. HALL. *Theory of Groups*. New York: Macmillan, 1959.
2. A. G. KUROSH. *Theory of Groups*, vols. 1 and 2. New York: Chelsea, 1958.
3. J. J. ROTMAN. *Theory of Groups*. Boston: Allyn and Bacon, 1965.
4. E. SCHENKMAN. *Group Theory*. New York: Van Nostrand, 1965.
5. W. R. SCOTT. *Group Theory*. Englewood Cliffs, New Jersey: Prentice-Hall, 1964.
6. G. HIGMAN, B. H. NEUMANN, and H. NEUMANN. "Embedding Theorems for Groups," *J. Lond. Math. Soc.*, 24 (1949), pp. 247–254.
7. H. WIELANDT. *Finite Permutation Groups*. New York: Academic Press, 1964.
8. D. S. ROBINSON. "On the Theory of Subnormal Subgroups," *Math. Z.*, 89 (1965), pp. 30–51.
9. H. ZASSENHAUS. *Theory of Groups* (2nd ed.). New York: Chelsea, 1958.
10. W. FEIT and J. G. THOMPSON. "Solvability of Groups of Odd Order," *Pacific J. Math.*, 13 (1963), pp. 775–1029.
11. P. HALL. "Frattini Subgroups of Finitely Generated Groups," *Proc. Lond. Math. Soc.* (3), 11 (1961), pp. 327–352.
12. A. I. MAL'CEV. "On Some Classes of Infinite Solvable Groups," *Amer. Math. Soc. Translations* (2), 2 (1956), pp. 1–22.
13. P. HALL and G. HIGMAN. "The p-Length of a p-Soluble Group and Reduction Theorems for Burnside's Problem," *Proc. Lond. Math. Soc.* (3), 6 (1956), pp. 1–42.

14. B. Huppert. "Lineare Auflösbare Gruppen," *Math. Z.*, 67 (1957), pp. 479–518.
15. B. Huppert. *Endliche Gruppen*, vols. 1 and 2. Berlin: Springer, 1967.
16. C. W. Curtis and I. Reiner. *Representation Theory of Finite Groups and Associative Algebras*. New York: Interscience, 1962.
17. M. Burrows. *Representation Theory of Finite Groups*. New York: Academic Press, 1965.
18. W. Burnside. *Theory of Groups of Finite Order* (2nd ed. reprint). New York: Dover, 1955.
19. O. Ore. "Contributions to the Theory of Groups of Finite Order," *Duke Math. J.*, 5 (1939), pp. 431–460.
20. W. R. Scott. "On a Result of B. H. Neumann," *Math. Z.*, 66 (1956), p. 240.
21. B. H. Neumann. "Groups with Finite Classes of Conjugate Subgroups," *Math. Z.*, 63 (1955), pp. 76–96.
22. R. Baer. "Finiteness Properties of Groups," *Duke Math. J.*, 15 (1948), pp. 1021–1032.
23. H. Schwerdtfeger. "Über eine Spezielle Klasse Frobeniusscher Gruppen," *Arch. Math.*, 13 (1962), pp. 283–289.
24. O. Zariski and P. Samuel. *Commutative Algebra*, vol. 1. Princeton, New Jersey: Van Nostrand, 1958.
25. B. H. Neumann. "Lectures on Topics in the Theory of Infinite Groups." Mimeographed notes (Tata Institute of Fundamental Research, Bombay, 1960).
26. H. Zassenhaus. "A Group-Theoretic Proof of a Theorem of MacLagan-Wedderburn," *Proc. Glasgow Math. Assoc.*, 1 (1952), pp. 53–63.
27. R. Brauer. "Representations of Finite Groups," in *Lectures on Modern Mathematics*, vol. 1 (Saaty, ed.). New York: Wiley, 1963.
28. W. Krull. "Über p-Untergruppen Endlicher Gruppen," *Arch. Math.*, 12 (1961), pp. 1–6.
29. I. Connell. "Problem 69," *Canad. Math. Bull.*, 6 (1963), p. 280 and 7 (1964), p. 145.
30. S. Picard. "Sur les Bases du Groupe Symmetrique et du Groupe Alternant," *Math. Ann.*, 116 (1939), pp. 752–767.
31. L. G. Kovács, B. H. Neumann, and H. de Vries. "Some Sylow Subgroups," *Proc. Royal Soc.*, A 260 (1961), pp. 304–316.
32. J. G. Thompson. "Finite Groups with Fixed-Point-Free Automorphisms of Prime Order," *Proc. Nat. Acad. Sci. U.S.A.*, 45 (1959), pp. 578–581.
33. A. Speiser. *Die Theorie der Gruppen von Endlicher Ordnung*. Berlin, 1937.
34. J. S. Brodkey. "A Note on Finite Groups with an Abelian Sylow Group," *Proc. Amer. Math. Soc.*, 14 (1963), pp. 132–133.
35. R. Baer. "Das Hyperzentrum einer Gruppe III," *Math. Z.*, 59 (1953), pp. 299–338.
36. W. R. Scott. "On the Order of the Automorphism Group of a Finite Group," *Proc. Amer. Math. Soc.*, 5 (1954), pp. 23–24.
37. J. C. Howarth. "Some Automorphisms of Finite Nilpotent Groups," *Proc. Glasgow Math. Assoc.*, 4 (1960), pp. 204–207.
38. E. Artin. *Geometric Algebra*. New York: Interscience, 1957.
39. L. E. Dickson. *Linear Groups* (reprint). New York: Dover, 1958.
40. H. Weyl. *The Classical Groups*. New Jersey: Princeton University Press, 1946.

41. J. Dieudonné. *Sur les Groupes Classiques*. Paris: Hermann, 1948.
42. R. W. Carter. "Simple Groups and Simple Lie Algebras," *J. Lond. Math. Soc.*, 40 (1965), pp. 193–240.
43. H. Wielandt. "Eine Verallgemeinerung der Invarianten Untergruppen," *Math. Z.*, 45 (1939), pp. 209–244.
44. H. Heineken. "Eine Bermerkung über Engelsche Elemente," *Arch. Math.*, 11 (1960), p. 321.
45. J. D. Dixon. "Complete Reducibility of Infinite Groups," *Canad. J. Math.*, 16 (1964), pp. 267–274.
46. E. Schenkman. "A Generalization of the Central Elements of a Group," *Pacific J. Math.*, 3 (1953), pp. 501–504.
47. K. W. Gruenberg. "The Engel Structure of Linear Groups," *J. Algebra*, 3 (1966), pp. 291–303.
48. H. Heineken. "Über ein Levisches Nilpotentzkriterium," *Arch. Math.*, 12 (1961), pp. 176–178.
49. B. I. Plotkin. "Generalized Soluble and Generalized Nilpotent Groups," *Amer. Math. Soc. Translations* (2), 17 (1961), pp. 29–116.
50. M. Rosenlicht. "On a Result of Baer," *Proc. Amer. Math. Soc.*, 13 (1962), pp. 99–101.
51. J. A. H. Sheppard and J. Wiegold. "Transitive Permutation Groups and Groups with Finite Derived Groups," *Math. Z.*, 81 (1963), pp. 279–285.
52. I. D. MacDonald. "Some Explicit Bounds in Groups with Finite Derived Groups," *Proc. Lond. Math. Soc.* (3), 11 (1961), pp. 23–56.
53. N. Blackburn. "Generalizations of Certain Elementary Theorems on *p*-Groups," *Proc. Lond. Math. Soc.* (3), 11 (1961), pp. 1–22.
54. R. D. Carmichael. *Groups of Finite Order* (reprint). New York: Dover, 1956.
55. G. A. Miller, H. F. Blichfeldt, and L. E. Dickson. *Theory and Applications of Finite Groups* (reprint). New York: Dover, 1961.
56. V. Dlab. "A Note on Powers of a Group," *Acta Sci. Math.* (*Szeged*), 25 (1964), pp. 177–178.
57. R. Baer. "Nilpotent Groups and Their Generalization," *Trans. Amer. Math. Soc.*, 47 (1940), pp. 393–434.
58. K. A. Hirsch. "Eine Kennzeichnende Eigenschaft Nilpotenter Gruppen," *Math. Nachr.*, 4 (1950), pp. 47–49.
59. P. Hall. "A Contribution to the Theory of Groups of Prime Power Orders," *Proc. Lond. Math. Soc.* (2), 36 (1933), pp. 29–95.
60. P. Hall. "Some Sufficient Conditions for a Group to be Nilpotent," *Illinois J. Math.*, 2 (1958), pp. 787–801.
61. H. Fitting. "Beiträge zur Theorie der Gruppen Endlicher Ordnung," *Jber. Deutsch. Math.-Verein.*, 48 (1938), pp. 77–141.
62. R. Baer. "Supersoluble Groups," *Proc. Amer. Math. Soc.*, 6 (1955), pp. 16–32.
63. C. R. Hobby. "The Frattini Subgroup of a *p*-Group," *Pacific J. Math.*, 10 (1960), pp. 209–212.
64. J. Szep and N. Itô. "Über die Faktorisation von Gruppen," *Acta Sci. Math.* (*Szeged*), 16 (1955), pp. 229–231.
65. K. Iwasawa. "Über die Struktur der Endlichen Gruppen deren Echte Untergruppen Sämtlich Nilpotent sind," *Proc. Phys. Math. Soc. Japan*, 23 (1941), pp. 1–4.

66. H. WIELANDT and B. HUPPERT. "Arithmetic and Normal Subgroup Structure," in *Proc. Symposia in Pure Math.*, vol. 6. Rhode Island: Amer. Math. Soc., 1962.
67. J. D. DIXON. "Complements of Normal Subgroups in Infinite Groups," *Proc. Lond. Math. Soc.* (3), 17 (1967) pp. 431–446.
68. P. HALL. "Nilpotent Groups." Mimeographed notes (Lectures to the Summer Seminar of the Canadian Mathematical Congress, 1957).
69. R. KOCHENDÖRFFER. "Hallgruppen mit Ausgezeichnetem Repräsentantensystem," *Acta Sci. Math. (Szeged)*, 21 (1960), pp. 218–223.
70. M. SUZUKI. "On the Existence of a Hall Normal Subgroup," *J. Math. Soc. Japan*, 15 (1963), pp. 387–391.
71. P. HALL. "On the Sylow Systems of a Soluble Group," *Proc. Lond. Math. Soc.* (2), 43 (1937), pp. 316–323.
72. R. W. CARTER. "Nilpotent Self-Normalizing Subgroups of Solvable Groups," *Math. Z.*, 75 (1961), pp. 136–139.
73. I. N. HERSTEIN. "A Remark on Finite Groups," *Proc. Amer. Math. Soc.*, 9 (1958), pp. 255–257.
74. J. G. THOMPSON. "Normal p-Complements for Finite Groups," *J. Algebra*, 1 (1964), pp. 43–46.
75. V. A. BELONOGOV. "Finite Groups with a Pair of Nonconjugate Nilpotent Maximal Subgroups," *Soviet Math.*, 6 (1965), pp. 566–567.
76. W. GASCHÜTZ. "Über die Φ-Untergruppen Endlicher Gruppen," *Math. Z.*, 58 (1953), pp. 160–170.
77. R. BAER. "Direkte Faktoren Endlicher Gruppen," *J. Reine Angew. Math.*, 192 (1953), pp. 167–179.
78. Z. JANKO. "Eine Bermerkung Über die Φ-Untergruppen Endlicher Gruppen," *Acta Sci. Math. (Szeged)*, 23 (1962), pp. 247–248.
79. V. DLAB and V. KOŘINEK. "The Frattini Subgroup of a Direct Product of Groups," *Czech. Math. J.*, 10 (1960), pp. 350–358.
80. J. L. ALPERIN. "Large Abelian Subgroups of p-Groups," *Trans. Amer. Math. Soc.*, 117 (1965), pp. 10–20.
81. W. E. DESKINS. "On Quasinormal Subgroups of Finite Groups," *Math. Z.*, 82 (1963), pp. 125–132.
82. N. ITÔ. "Über das Produkt von Zwei Abelschen Gruppen," *Math. Z.*, 63 (1955), pp. 400–401.
83. H. WIELANDT. "Über die Normalstruktur von Mehrfach Faktorisierten Gruppen," *J. Austral. Math. Soc.*, 1 (1960), pp. 143–146.
84. O. KEGEL. "Zur Struktur Mehrfach Faktorisierter Endlicher Gruppen," *Math. Z.*, 87 (1965), pp. 42–48.
85. M. SUZUKI. *Structure of a Group and the Structure of its Lattice of Subgroups.* Berlin: Springer, 1956.
86. P. HALL. "Complemented Groups," *J. Lond. Math. Soc.*, 12 (1937), pp. 201–204.
87. A. H. CLIFFORD. "Representations Induced in an Invariant Subgroup," *Ann. of Math.*, 38 (1937), pp. 533–550.
88. H. ZASSENHAUS. "Über Endliche Fastkörper," *Abh. Math. Sem. Uni. Hamburg*, 11 (1936), pp. 185–220.
89. N. ITÔ. "A Note on A-Groups," *Nagoya Math. J.*, 4 (1952), pp. 79–81.
90. B. HUPPERT. "Monomiale Darstellung Endlicher Gruppen," *Nagoya Math. J.*, 6 (1953), pp. 93–94.

91. D. Suprunenko. "Soluble and Nilpotent Linear Groups," *Translations of Math. Monographs*, vol. 9. Rhode Island: Amer. Math. Soc., 1963.
92. H. Zassenhaus. "Beweis eines Satzes über Diskrete Gruppen," *Abh. Math. Sem. Uni. Hamburg*, 12 (1938), pp. 289–312.
93. R. C. Thompson. "Commutators in the Special and General Linear Groups," *Trans. Amer. Math. Soc.*, 101 (1961), pp. 16–33.
94. K. Goldberg. "Unimodular Matrices of Order 2 that Commute," *J. Wash. Acad. Sci.*, 46 (1956), pp. 337–338.
95. E. C. Dade. "Abelian Groups of Unimodular Matrices," *Illinois J. Math.*, 3 (1959), pp. 11–27.
96. E. C. Dade. "Answer to a Question of R. Brauer," *J. Algebra*, 1 (1964), pp. 1–4.
97. L. Weisner. "Condition that a Finite Group is Multiply Isomorphic with Each of its Irreducible Representations," *Amer. J. Math.*, 61 (1939), pp. 709–712.
98. R. Brauer. "A Note on Theorems of Burnside and Blichfeldt," *Proc. Amer. Math. Soc.*, 15 (1964), pp. 31–34.
99. R. Brauer and J. Tate. "On Characters of Finite Groups," *Ann. of Math.*, 62 (1955), pp. 1–7.
100. I. M. Isaacs and D. S. Passman. "Groups with Representations of Bounded Degree," *Canad. J. Math.*, 16 (1964), pp. 299–309.
101. W. Burnside. "On Criteria for the Finiteness of the Order of a Group of Linear Substitutions," *Proc. Lond. Math. Soc.* (2), 3 (1905), pp. 435–440.
102. J. E. Roseblade. "On Certain Classes of Locally Soluble Groups," *Proc. Cambridge Philos. Soc.*, 58 (1962), pp. 185–195.

Index of Terms